This authoritative volume is devoted to considerations of the different modifications the molecular biochemical units or systems may undergo in the course of descent along the phylogenic pathways of organisms.

It presents a critical survey of the findings available in the literature concerning molecular phylogeny, one of the fundamentals of molecular biology, which, from certain standpoints, can be said to represent the evolution of biochemical processes in nature.

Although the author's approach is an organismic one the collection of data acquired so far has been made possible through the recent development of molecular biology, which brings in the possibility of a molecular approach to the evolution of organisms as recorded in the phylogenic tree.

This valuable volume, the work of a world-renowned expert, will be of interest to biologists, biochemists, molecular biologists, biophysicists and students of evolution in general. It also contains information of value to chemists, ecologists and systematists.

A MOLECULAR APPROACH TO PHYLOGENY

A MOLECULAR APPROACH
TO PHYLOGENY

by

MARCEL FLORKIN

Professor of Biochemistry, University of Liège (Belgium)

ELSEVIER PUBLISHING COMPANY

AMSTERDAM / LONDON / NEW YORK

1966

ELSEVIER PUBLISHING COMPANY
335 JAN VAN GALENSTRAAT, P.O. BOX 211, AMSTERDAM

AMERICAN ELSEVIER PUBLISHING COMPANY, INC.
52 VANDERBILT AVENUE, NEW YORK, N.Y. 10017

ELSEVIER PUBLISHING COMPANY LIMITED
RIPPLESIDE COMMERCIAL ESTATE, RIPPLEROAD, BARKING, ESSEX

LIBRARY OF CONGRESS CATALOG CARD NUMBER 66-13574

WITH 52 ILLUSTRATIONS AND 20 TABLES

PRINTED IN THE NETHERLANDS

Preface

In a previous work (*L'évolution biochimique*, Masson, Paris, 1944) the author presented evidence for the existence of modifications at the molecular level in the course of descent along the phylogenic pathways of organisms. This thesis has been widely accepted.

Since the advent of molecular biology, the search for evolutionary aspects at the molecular level as well as at the level of organisms has gained firmer ground and it is now possible to consider molecular evolution with more precision, and to attempt to distinguish its different aspects in parallel with the phylogeny of organisms.

The subject matter of this book was originally the topic of seminars presented by the author at the Friday Harbor Marine Laboratory (Washington, U.S.A.) during the summer of 1963, at the Department of Biophysics at Rio de Janeiro, during the summer of 1964 and at the Marine Biological Laboratory of Roscoff (France) during the summer of 1965. To all of those who took part in the discussions, the author expresses his gratitude.

Liège, April 1966   MARCEL FLORKIN

Contents

Preface V

Chapter 1. Introduction 1

Chapter 2. Basic concepts 5

Chapter 3. Phylogeny of peptides and proteins 11

Chapter 4. Biosynthesis and phylogeny. 34

Chapter 5. Chitin, chitinolysis and phylogeny 50

Chapter 6. Terminal products of nitrogen metabolism, ecological and phylogenic aspects 62

Chapter 7. Hemolymph osmotic effectors in insect phylogeny 89

Chapter 8. Definition, in terms of molecular evolution, of special characteristics in the carbohydrate metabolism of insects 115

Chapter 9. Paleoproteins 133

Chapter 10. Evolving organisms and molecules 157

Subject index 166

Chapter 1

Introduction

Since the beginning of the 19th century, the comparative viewpoint has increasingly occupied naturalists interested in the dynamic aspects of life. Since then also, the tendency has been towards the identification of aspects common to all forms of life. This tendency has been preeminently represented by Gerdy in France, and by Tiedemann in Germany. It contrasts sharply with the collection of data related to the infinite diversity of organisms and to the multiple aspects they have assumed in the course of evolution. This systematic search for unity may be said to have had its origin in the works of Bichat, whose thoughts were oriented towards the formulation of a general anatomy, which was to be born when Schwann later formulated his cell theory.

The opposing tendency originated in Etienne Geoffroy Saint-Hilaire's school. The famous initiator of *anatomic philosophy* was in search of a description of the physiological characteristics of the greatest possible number of organisms and it is this tendency which is illustrated in the *Traité de Physiologie Comparée de l'Homme et des Animaux* published in 1838 by a professor of Montpellier, Antoine Dugès. It is this same tendency that we find in the monumental *Leçons sur la Physiologie et l'Anatomie Comparée* by Milne-Edwards, fourteen volumes of which were printed between 1857 and 1880.

When Darwin published his celebrated book in 1859, the accent was laid on the fundamental genetic unity of organisms and this favored agreement on Claude Bernard's definition of general physiology as the study of phenomena common to all organisms, or as the physics and chemistry of living cells. The general cellular phenomena are obviously of paramount importance, and it is not surprising that the physiological and molecular aspects common to all cells have since aroused such constant interest.

On the other hand, the traits of the different categories or organisms were considered by the comparative physiologists under their phenomenological aspects, as being "functions" which could be defined at the level of the organism as a whole, and within the framework of the relationship of the organism to its environment. Still influenced by the idea of *anatomia animata*, the physiologists identified at the organismic level the material substratum of each "function": vascular systems for the circulation of fluids, respiratory organs for ventilation, glands for secretion, etc. They identified the material object of the "function": a gaseous medium for ventilation, a liquid object for currents, etc. Classical physiologists, having identified the mill and the grain, looked for the wind acting on the mill and in each case described a "stimulus" responsible for initiating the reaction of the organ to its material object. Comparative physiology was therefore led to compare analogous systems without any reference to homologies, which were left to the realm of comparative anatomy.

Unfortunately this non-phylogenic characteristic of comparative physiology was reflected in the first efforts of comparative biochemists. When Otto von Fürth published his *Vergleichende Chemische Physiologie der Niederen Tiere* in 1903, he was strongly under the influence of the physiological way of thinking, and he grouped his collection of data on invertebrates in physiological divisions: blood, respiration, nutrition, excretion, animal poisons, secretions, muscles, supporting structures, etc. Under each of these headings, the phenomenological aspects presented by the different phyla are successively considered. In fact, it is the physiological notion of "function", residing in the relationship of substratum to object, which prevails in Von Fürth's book, devoted as it is to the comparison of similar phenomena, without reference to phylogeny. The same approach is found in Baldwin's *An Introduction to Comparative Biochemistry*, the first edition of which appeared in 1937. "The task of the biochemist is, after all, the study of the physicochemical processes associated with the manifestations of what we call life— not the life of some particular animal or group of animals, but life in its most general sense", writes Baldwin. When, however, we are

Elsevier's Scientific Publications

For information about new books in the following fields, please check square(s) and complete reverse of this card.

- ☐ PHYSICAL AND THEORETICAL CHEMISTRY
- ☐ ORGANIC CHEMISTRY
- ☐ INORGANIC CHEMISTRY
- ☐ ANALYTICAL CHEMISTRY
- ☐ BIOLOGY
- ☐ SUGAR PUBLICATIONS
- ☐ BIOCHEMISTRY
- ☐ BIOPHYSICS
- ☐ CLINICAL CHEMISTRY
- ☐ PHARMACOLOGY
- ☐ TOXICOLOGY
- ☐ PSYCHIATRY
- ☐ NEUROLOGY
- ☐ ATHEROSCLEROSIS

(please print or type)

Name: ....

Address: ....

....

....

....

Elsevier's Scientific Publications

You received this card in one of our publications. It would greatly assist us in serving you further if, when returning it for more information, you would indicate below how you heard of the book or books now in your possession. We thank you for your co-operation.

- ☐ Bookseller's recommendation
- ☐ Books sent on approval by bookseller
- ☐ Displays in bookshops
- ☐ Reviews
- ☐ Advertisements
- ☐ Personal recommendation
- ☐ References in books and journals
- ☐ Publisher's catalogue
- ☐ Circular received from publisher
- ☐ Circular received from bookseller
- ☐ Listing in a subject catalogue of bookseller

POSTCARD

ELSEVIER PUBLISHING COMPANY

P.O. BOX 211

AMSTERDAM-W.
THE NETHERLANDS

concerned with systematics and phylogeny, it is, of course, precisely to particular groups of animals that our interest is directed. No reference is made to the phylogeny of molecules in Baldwin's *Introduction*. To be sure, he does make tentative use of biochemistry to unravel aspects of animal phylogeny, but this is an entirely different methodological approach. Baldwin's book, which had the merit of arousing a great deal of interest in comparative biochemistry, was based essentially on the categories of physiological chemistry, as Von Fürth's book had been.

In 1926, the great microbiologist Kluyver and his collaborator Donker published a paper entitled *Die Einheit in der Biochemie*, in which they formulated a principle, the fruitfulness of which has been constantly confirmed. The central postulate of this principle is that each step in a metabolic process can be reduced to a number of dehydrogenations and hydrogenations. Kluyver's conception allows the formulation of a hypothesis when considering a metabolic step of unknown nature, but one involving a metabolite the nature of whose participation in another known biochemical system is already understood. The fruitfulness of this principle has been particularly well illustrated in the brilliant studies of Van Niel on photosynthesis. Van Niel demonstrated the photosynthetic nature of the metabolism of green sulfur bacteria by showing that these bacteria perform the dehydrogenation of hydrogen sulfide and utilize carbonic anhydride as hydrogen acceptor. This results in the formation of organic molecules with the liberation of sulfur. Applying Kluyver's principle to this case, Van Niel conceived photosynthesis as the dehydrogenation of water and the hydrogenation of carbonic anhydride with the liberation of oxygen and the formation of an organic molecule. From that time onwards, photosynthesis ceased to be regarded as a molecular rearrangement of carbonic acid combined with chlorophyll with the formation of formaldehyde peroxide, and was included in the ranks of dehydrogenation and hydrogenation systems. Comparative biochemistry, according to Kluyver, does not deal with phylogeny or evolution, but is concerned, directly, with the study of the unity of biochemical systems corresponding to the same metabolic step in different organisms and, reciprocally, with the dynamics of known biochemical systems which may serve to explain a similar metabolic step in other, even phylogenetically unrelated, organisms. Since Kluyver's principle was formulated, we have learnt the biochemical similarity of subunits of cells, such as mitochondria, or endoplasmic reticulum, present in cells in general.

In a previous publication, *L'Évolution Biochimique* (Masson, Paris, 1944), the author developed the concept of molecular changes along the phylogenic tree. This notion has generally been accepted. It has taken on new aspects in the light of current thinking about the role of deoxyribonucleic acid in hereditary information transfer.

The present volume is devoted to a consideration of the different modifications the molecular biochemical units or systems may undergo in the course of descent along the phylogenic pathways of organisms. It has been the subject of a number of seminars held at the University of Brazil (Rio de Janeiro) in the Department of Biophysics, at the instigation of Professor Carlos Chagas. The author received many fruitful suggestions during the discussions which took place at these seminars and wishes to thank those who took part for their kind collaboration.

Chapter 2

Basic Concepts

Living organisms are made up of cells, of modified cells and of the products of cell activities. Each cell is a "constellation" of macro-molecules, molecules and ions assembled according to a common scheme, this scheme also prevailing in monocellular organisms. Not only the architecture of this scheme, but also the nature of its constituents, show a remarkable degree of similarity from cell to cell. This similarity is the reality behind the concept of the "unity of life". In this context, the "unity of biochemical plan" is nothing more than the cell theory stated in plain chemical terms. As already proposed by the founder of the cell theory, Theodor Schwann, in his epoch-making little book published in 1839, in the same way as they are *units of structure*, cells are *units of metabolism*. The unity of structure and metabolism in all living beings is an expression of cellular continuity, and of the persistence, through this continuity, of a collection of definite sequences of purine and pyrimidine bases, these sequences controlling the biosynthesis of the collection of enzymes found in each cell. In spite of a certain degree of unity in the nature of at least a part of this collection, no cell is limited to the underlying chemical similarity, which is only the canvas on which the cells have embroidered their differentiations, the "unity of plan" remaining an abstraction in the human mind. The increasingly molecular approach to biology brings us to consider a species as consisting of groups of individuals with more or less similar combinations of sequences of purine and pyrimidine bases in their macro-molecules of DNA, and with a system of operators, controllers and repressors leading to the biosynthesis of similar sequences of amino acids, the integration of which, in one cell, or in a number of variably differentiated cells, leads to similar structural and functional characteristics, adapted to the ecological niche in which the species flourishes.

Bibliography p. 10

In order to avoid misunderstandings, we should agree, at the start, on the definition of some of the basic concepts of comparative biochemistry. The biochemical compounds, molecules or macromolecules, which show signs of chemical kinship, we shall call *isologues*. Cytochrome, peroxidase, catalase, hemoglobin and chlorocruorin exhibit this isology, as they are heme derivatives. In the case of the hemoglobins of two men who are identical twins, the maximum degree of isology obtains. It is less pronounced when we consider the hemoglobins of a dog and of a jackal, and still less if we consider those of a dog and of a horse. In these cases, the protoheme is identical, but the degree of isology depends on the structure of globin, the protein moiety of the hemoglobin macromolecule. If we compare a hemoglobin and a cytochrome *c*, the isology obtains only at the level of the heme moiety, the sequences of amino acids in the protein part being non-isologous. Isology is a chemical concept.

Extensive studies on the primary structure of a variety of protein macromolecules (see Vegotsky and Fox, 1962), and the knowledge that this primary structure stands in a definite relationship to the sequences of bases of DNA indirectly controlling their biosynthesis, have led to the conclusion, among others, that the "insulin-determining" sequences of bases of the pig and of the sperm whale are

```
                              ┌─Heme─┐
 – Gly – – – Lys·Gly – – – Phe – – – CyS – – CyS·His·Thr·Val·Glu –
              10                                    20

 Gly·Gly – His·Lys – Gly·Pro·Asn·Leu – Gly – Phe·Gly·Arg – – Gly·Gln·Ala –
              30                                    40

 Gly – – Tyr·Thr – Ala·Asn – – Lys – – – Try – Glu – – – – – Tyr·
              50                          60

 Leu – Asn·Pro·Lys·Lys·Tyr·Ileu·Pro·Gly·Thr·Lys·Met·Ileu·Phe – Gly –
              70                                    80

 Lys·Lys – – – Arg – Asp·Leu – – Tyr·Leu·Lys·Lys – – – – COOH
          90                          100
```

Fig. 1. Identical residues in the amino acid sequences of all presently known vertebrates and yeast cytochrome *c*. Numbering of the residue positions follows that of the vertebrate proteins; yeast cytochrome *c* has five extra residues at the N-terminal end and one residue less at the C-terminal end. The 58 identical residues in all species are given and the non-identical residues are indicated by a dash. (Smith and Margoliash, 1964)

identical. Studies on ACTH, ribonuclease, melanotropic hormone, insulin, cytochrome c, hemoglobin, etc., have shown that when each of these macromolecules is obtained from different organisms, many similarities in primary structure are observed. Fig. 1 illustrates the case of cytochrome c. Such a degree of isology is incompatible with chance effects*, and points to the persistence, through the whole of the evolutionary tree, of very ancient base sequences of DNA. These sequences, reproduced through the ages with seemingly unwavering constancy, may be called *homologues* in the sense used by the biologist, *i.e.*, connoting a common origin and a common line of descent. *Homology*, considered at the molecular level, is thus not a chemical concept, but a genetic concept, in contrast to the use of the word in enzymology, where it means "a similar function". In the concept of homology as presented here, properties (such as enzyme activity, hormone activity, etc.) are irrelevant, and the emphasis falls exclusively on the common descent, by replication of a DNA sequence of bases, of a common ancestor. The whole of our present knowledge suggests the great probability that the very isologous primary structures of proteins are replicas of very isologous base sequences in nucleic acids and that, like these sequences of bases, they may be qualified as homologous. The term may likewise be applied to chains of homologous protein biocatalysts, and also to the results of a biosynthesis catalyzed by a homologous enzyme chain. This usage makes clear the distinction between isology and homology. ATP is isologous in all cells, but it is not always homologous, being, for instance, the product of the action of one chain of biocatalysts in glycolysis, and of another in oxidative phosphorylation. Bile acids, on the other hand, are homologous in all vertebrates, as they are biosynthetized by pathways catalyzed by homologous enzyme chains.

---

* This statement, of course, connotes a certain degree of postulation and we must await more detailed information on primary structures in order to be able to draw up a theory of the relationships between homology and the degree of isology. On the other hand, the usefulness of a marked degree of isology in detecting homology may be limited to a short section of the phylogenic tree.

TABLE I

(McElroy and Seliger, 1963)

Organism	Nature of reactants	Peak light emission
Luminous bacteria	$FMNH_2 + RCHO + O_2 + E$	495 mμ
Fireflies	$LH_2 + ATP + Mg^{2+} + O_2 + E$	562 mμ
Cypridina (crustacean)	$LH_2 + O_2 + E$	460 mμ
Odontosyllis (polychaete worm)	$LH_2 + O_2 + E$	510 mμ
Pholas dactylus (luminous clam)	$NADH + FMN + O_2 + E$	blue
Omphali flavida (fungus)	$NADH + X + O_2 + E$	530 mμ
Renilla reniformis (sea pansy)	$LH_2 + AMP + O_2 + E$	blue
Gonyaulax polyhedra (protozoan)	$LH_2 + E + O_2$	blue
Apogon (fish)	$LH_2 + E + O_2$	460 mμ

The term *analogous* is applied to biochemical units which play the same role in different biochemical systems. The luciferins listed in Table I are analogues, though they are not isologues.

A most interesting case of analogy is provided by the different kinds of oxygen-carriers: hemoglobins, chlorocruorins, hemocyanins and hemerythrins. Another most interesting case is provided by the proteins of a parasite and of its host, these proteins being analogous in their action as antigens. Analogy, as we have said, refers to structures presenting similar properties: enzymatic activity, reversible oxygenation, etc. On the other hand, if a common primary structure, *e.g.* a sequence of amino acids, were discovered in different cells where it could be proved that the initial prototypes of these sequences were different (a most improbable event according

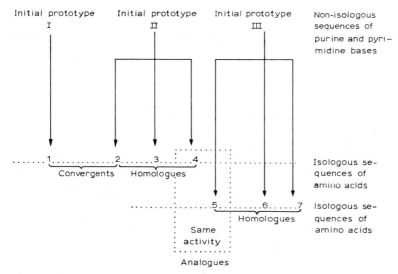

Fig. 2. Isology, homology, analogy and convergence. The roman figures designate sequences of purine and pyrimidine bases. The arabic figures designate sequences of amino acids. (Florkin, 1962)

to our present theories), we ought to consider this a convergence.

The concepts of isology, homology, analogy and convergence can be most clearly differentiated when applied to the study of primary protein structures, *viz.* of amino acid sequences and of their initial prototypes made up of sequences of purine and pyrimidine bases in nucleic acids (Fig. 2).

In the history of comparative physiology, homologous and analogous components have not been distinguished. In the present study, this fundamental distinction may be usefully preserved as an aid to sounder appreciation of the adaptive nature of molecular evolution and of the site of impact of natural selection.

BIBLIOGRAPHY

FLORKIN, M., Isologie, homologie, analogie et convergence en biochimie comparée. *Bull. Classe Sci., Acad. Roy. Belg.*, 48 (1962) 819–824.

McELROY, W. D., AND H. H. SELIGER, Origin and evolution of bioluminescence, in: A. I. OPARIN (Ed.), *Evolutionary Biochemistry*, Pergamon, Oxford, 1963, pp. 158–168.

SCHWANN, TH., *Mikroskopische Untersuchungen über die Übereinstimmung in der Struktur und dem Wachstum der Thiere und Pflanzen*, Sanderschen Buchhandl., Berlin, 1839.

SMITH, F. L., AND E. MARGOLIASH, Evolution of cytochrome *c*. *Federation Proc.*, 23 (1964) 1243–1257.

VEGOTSKY, A., AND S. W. FOX, Protein molecules: intraspecific and interspecific variations, in: M. FLORKIN AND H. S. MASON (Eds.), *Comparative Biochemistry*, Vol. IV, Academic Press, New York, 1962, pp. 185–244.

Chapter 3

Phylogeny of Peptides and Proteins

When, in Chapter 1, we defined homology in the sense adopted by comparative biochemists (Florkin, 1962; 1963 a and b; 1964; 1965), we underlined the fact that the primary structure of a protein, *viz.* the sequence of the amino acids composing a chain in the protein, is an indirect reflection of the sequence of purine and pyrimidine bases in the DNA controlling the synthesis of this protein. Regarded in this way, the sequences of amino acids may be considered homologous if they show a high degree of isology. The proteins whose chain sequences have so far been determined are in general substances having the same properties in different organisms: insulin, adrenocorticotropic hormone (ACTH) of the pituitary body, ribonuclease, melanotropic hormone, cytochrome *c*, hemoglobin, etc. Those proteins exhibiting the *same* properties in different organisms were also found to display the high degree of isology by which we detect homology, though they show some differences in composition according to the position of the organism in the branches of the phylogenic tree.

What do we understand by the phylogeny of a protein? We are concerned with the phylogeny of a protein when we establish comparison between homologous forms of this protein at different levels of the same phyletic series of the organisms synthetizing these forms.

Consider, for example, chlorocruorin: the oxygen carrier in the flabelligeridae (chlorhemian spioniforms), sabellidae and serpulidae. This carrier is present in the blood of those three families of polychete annelids. The chlorhemians are sedentary polychete annelids, which are descended from the errant polychete annelids: those forms whose preoral lobe is not sunk into the first segment of the metasome, and which feed on floating plankton gathered by means of posterior antennae in the form of long palps bearing a ciliated gutter. They live in sand or mud and secrete a membranous tube

Bibliography p. 32

covered with a fine layer of slime. The chlorhemians are spioniforms which have lost the dissepiments and even the external segmentation. Their blood is green, and their palps are folded forward. Related to the spioniforms are the cryptocephalic annelids, having a preoral lobe sunk into the first segment of the trunk, but possessing furrowed appendages like those of the spioniforms. Sedentary and tubicolous, the cryptocephalae comprise two sub-divisions: the sabellariides, which although sedentary and microphagic, have retained an un-even number of antennae, and the sabelliforms which have an even number of antennae, and palps forming a multicolored corolla. The sabelliforms are divided into the sabellidae, having a mucous tube, membranous or cornified, and the serpulidae, possessing a calcare-ous tube.

The blood of spioniforms other than the chlorhemians is colored red by hemoglobin, as in *Sabellaria*. In the sabelliforms, sometimes called the serpuliforms, chlorocruorin is the characteristic green blood pigment. All the sabellidae so far studied contain it. Among the serpulidae, blood of species in the genus *Serpula* contains both chlorocruorin and hemoglobin, whilst in the genus *Spirorbis*, one species, *S. borealis*, has blood colored by chlorocruorin, another, *S. corrugatus*, contains hemoglobin, and a third, *S. militaris* has colorless blood. H. M. Fox (1949) did not find chlorocruorin in the tissues or in the coelomic fluid of the forms having chlorocruorin in the blood. No doubt, in those forms which contain it, the syn-thesis of chlorocruorin is a variant of hemoglobin synthesis as it was present in their annelid ancestors possessing this synthetic mechanism. Also, chlorocruorin is a close isologue of annelid hemo-globin and possesses many similar properties. The heme of chloro-cruorin, or chlorocruoroheme, differs from protoporphyrin in only one small detail, the oxidation of vinyl group 2. As for the protein portion, it is very similar to that found in annelid hemoglobin, as the data in Table II show.

In the case of chlorocruorin, we have a chemical entity very isologous to hemoglobin and present in divisions of our systematic classification, and the comparative morphology of these classes shows their phylogenic relation to other classes the blood of whose

TABLE II

(For literature, see Florkin, 1948)

	Iso-electric point	Molecular weight $\times$ 17 000 (+)	Containing amino acids			
			Cystine (%)	Arginine (%)	Histidine (%)	Lysine (%)
Hemoglobin of horse	6.78	4	0.74	3.57	8.13	8.31
Hemoglobin of Lombricus	5.28	192	1.41	10.07	4.68	1.73
Hemoglobin of Arenicola	4.76	192	4.08	10.04	4.03	1.85
Chlorocruorin of Spirographis	4.3	192	1.64	9.64	2.38	3.64

members contains hemoglobin. In this context, we may safely assume that there is homology between hemoglobin and chlorocruorin, and that the latter is biosynthetized in place of the former. In other words, it is an example of evolution with change of structure, at the level of the heme as well as of the globin.

In the present situation, we must adopt the methodical rule to be led by knowledge of phylogeny in our search for biochemical evolution, rather than to be brought by biochemistry to the discovery of new aspects of phylogeny. The Ariadne's thread of comparative biochemistry can only be the knowledge of phylogeny, in which is integrated the treasure of knowledge about living organisms accumulated by generations of naturalists. On the other hand, only homologous aspects of biochemistry can be of use.

Before discussing the phylogeny of proteins on the basis of data provided by comparative biochemistry, it is important to be informed on the relationship between the primary structure, *viz.* the amino acid sequence, genetically determined, and the properties of protein, for example as a catalyst. We have, in this respect, enough information in the cases of ribonuclease and of lysozyme.

Fig. 3 shows the sequence of amino acid residues in bovine pancreatic ribonuclease. It appears from the whole of our present

Bibliography p. 32

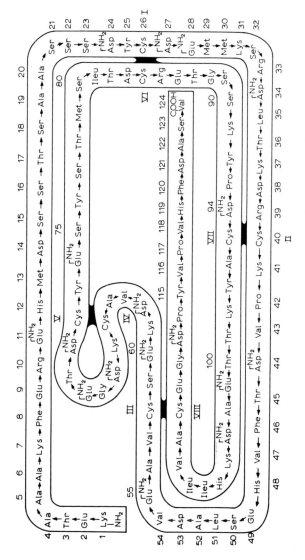

Fig. 3. Sequence of amino acid residues in bovine pancreatic ribonuclease.
(Smyth *et al.*, 1963)

knowledge that the relative positions of imidazole rings of the histidine residues 12 and 119 must be in a definite relative position in order to exert the proper "push–pull" effect that accelerates the cleavage of the specific type of phosphate ester bound in RNA. The positively charged site of the lysine residue at position 41 is also essential for activity, and it appears clear in this case that the primary structure of the protein is responsible for the tertiary structure which determines the topography of the active site. The same conclusion can be drawn from the study of lysozyme.

Such a view cannot, of course, be extended to all protein catalysts before a proper study is made of their primary structures. It can be stated, however, that in the two enzymes referred to, the information carried by the DNA and transferred in the cytoplasm through messenger RNA, or by other methods still unknown, controls the tertiary structure through the amino acid sequences, and that in these cases the catalytic function of the macromolecule is genetically determined.

That the transmission of information may be accompanied by changes is illustrated, for example, by the study of the somatotropins and of the insulins of sheep and sperm-whale, in which we find great similarities in primary structures in the insulins and striking differences in the somatotropins, an example of the fact that variation does not necessarily take place in all components of the transmitted information, or necessarily at the same speed. This fact led Anfinsen to propose the useful concept of the protein spectrum; "Since proteins can be modified without loss of function, it seems certain that the permissible degree of modification, in terms of fractions of their total structure, will vary somewhat from molecular species to species". As Anfinsen suggests, it does not seem too farfetched to think of the proteins of a given organism as being subdivisible into those that have structures quite closely tailored to an essential functional requirement, those that are designed with only moderate "efficiency" or whose function is relatively dispensable, and those that are intermediate. For instance, we cannot imagine a cell deprived of the enzymes of the first stages of glycolysis, while we know that certain human beings are completely deprived of serum albumin and that their clinical abnormality is only slight. According to this view, we may "express" a species in terms of a hierarchy of protein structures, ranging in "violability" from none at all to very much. During mutation and natural selection, one end of the spectrum of proteins would remain little changed, and the other end change considerably. The enzymes of glycolysis would be persistently reproduced, whilst serum albumin would markedly fluctuate.

This process explains the persistence of those proteins which we recognize as accounting for the unity of biochemistry. It also accounts for the evolution with change of molecular structure as well as for the appearance of new proteins in phylogeny. We are more and more ready to accept that each species is characterized by a unique assembly of proteins, but our knowledge of the primary structure of proteins is not advanced enough to permit very extensive speculation on their evolution.

If each species is characterized by a unique assembly of proteins, evolution with structural change ought to be explicable in relation to the protein specificity of each species.

Such hormones as growth hormone show, from fish to mammals, great variations in properties which are associated with specific antigenic reactions. This also applies to ACTH, another of the hormones of the pars distalis of the pituitary gland. It consists of 39 amino acid residues, and the structural sequence and composition differ in pig, sheep and man. But an interesting fact, pointed out by Harris (1960) and by Burgers (1961), is the close structural relationship which exists between these ACTH molecules and the melanocyte-stimulating hormone (MSH) which is secreted in the pars intermedia of the pituitary gland.

ACTH and MSH have in common the heptapeptide sequence Met·Glu·His·Phe·Arg·Try·Gly (Fig. 4). As ACTH is secreted by the pars distalis and MSH by the pars intermedia of the pituitary gland, both embryologically derived from Rathke's pouch, perhaps the possession of a common heptapeptide may be the consequence of this common embryological origin, and the differentiation of Rathke's pouch into two different glandular regions may be accompanied by the specialization of the same central heptapeptide into two different hormones.

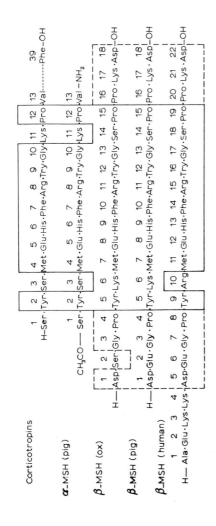

Fig. 4. Amino acid sequences in corticotropins and in the melanocyte-stimulating hormones. (Harris, 1960)

Isotocin (whiting, black pollack, hake, carp)

1	2	3	4	5	6	7	8	9

Cys – Tyr – *Ileu* – *Ser* – Asp(NH$_2$) – Cys – Pro – *Ileu* – Gly(NH$_2$)

Mesotocin (frog)

1	2	3	4	5	6	7	8	9

Cys – Tyr – *Ileu* – *Glu(NH$_2$)* – Asp(NH$_2$) – Cys – Pro – *Ileu* – Gly(NH$_2$)

Oxytocin (ox, pig, horse, sheep, whale, man, chicken)

1	2	3	4	5	6	7	8	9

Cys – Tyr – *Ileu* – *Glu(NH$_2$)* – Asp(NH$_2$) – Cys – Pro – *Leu* – Gly(NH$_2$)

Vasotocin (whiting, black pollack, hake, carp, frog, chicken)

1	2	3	4	5	6	7	8	9

Cys – Tyr – *Ileu* – *Glu(NH$_2$)* – Asp(NH$_2$) – Cys – Pro – *Arg* – Gly(NH$_2$)

Arginine–vasopressin (ox, horse, sheep, whale, man)

1	2	3	4	5	6	7	8	9

Cys – Tyr – *Phe* – *Glu(NH$_2$)* – Asp(NH$_2$) – Cys – Pro – *Arg* – Gly(NH$_2$)

Lysine–vasopressin (pig)

1	2	3	4	5	6	7	8	9

Cys – Tyr – *Phe* – *Glu(NH$_2$)* – Asp(NH$_2$) – Cys – Pro – *Lys* – Gly(NH$_2$)

Fig. 5. Hormones of the neurohypophysis. (For literature, see Acher, 1963; Acher, *et al.*, 1965a)

Fig. 5 shows the structure of the neurohypophyseal peptides in a series of animals. In fishes as well as in mammals, two peptides are found. They are isotocin and vasotocin in fishes (in the carp as well as in the marine fishes)*. In the mammals, the two peptides are oxytocin and vasopressin. In both cases, the peptides are constituted by a sequence of 9 amino acids, with a disulfide bond between 1 and

---

* Since this text was written Acher *et al.* (1965b) have isolated a new hormone, glumitocin (Ser$_4$–Gln$_8$–oxytocin) from the neurohypophysis of the skate *Raia clavata*.

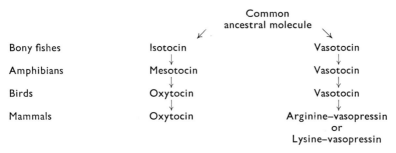

Fig. 6. Hypothetical phylogeny of neurohypophyseal peptides in vertebrates.
(Acher, 1963, modified)

6. Acher (1963) proposes that the isotocin of fishes should be considered the ancestor of the oxytocin of mammals, *i.e.* as having
appeared in a common ancestor of both, and in the same sense, that
vasotocin should be regarded as an ancestor of arginine–vasopressin. The appearance of the latter would have resulted from the
replacement, in position 3, of a residue of isoleucine by a residue of
phenylalanine. If it is accepted that isotocin is the ancestor of oxytocin, the modification of the former to the latter calls for two substi-
tutions only (in 4 and 8). In each category of vertebrates two homologous peptides are found. Their functions may, however, differ
as is, for instance, the case in vertebrates, where oxytocin is active
in reproduction and vasopressin in hydromineral regulation.

Acher (1963) represents the phylogeny of neurohypophyseal
peptides by the hypothetical scheme of Fig. 6, in which the possibility
is suggested of their derivation from a single ancestral molecule.
The two structural genes (if any) controlling the sequences of amino
acids in the two peptides could have resulted from the duplication
of an ancestral gene.

According to our present views, the evolutionary changes of
protein structure are indirect results of changes in DNA structure.
In the case of the four independent genes controlling the three normal
human hemoglobins, Ingram has suggested that the increase in
diversity of the hemoglobins is a result of an increase in the number
of genes. Hemoglobin A of man (adult hemoglobin) contains two
α- and two β-chains. The other human hemoglobins also contain

two α-chains and the overall formulation of each hemoglobin therefore begins with α_2^A.

$$\text{Hemoglobin A (adult)} = \alpha_2^A \beta_2^A$$

$$\text{Hemoglobin F (fetal)} = \alpha_2^A \gamma_2^F$$

$$\text{Hemoglobin A}_2 = \alpha_2^A \delta_2^A$$

The four chains are different with respect to their overall amino acid composition but, in the opinion of Ingram, a single myoglobin-like heme protein was their common forerunner. The scheme he proposes (Fig. 7) involves an increase in the number of hemoglobin genes from one to five, by duplication and translocation. At the stage of the single myoglobin-like forerunner, with a single heme group, and molecular weight of about 17 000, the heme protein inside the cells is assumed to have been the same as the circulating one. The three-dimensional arrangement in this heme protein was now subject to evolutionary change, but not drastically if it was to keep the property of reversible oxygenation. If we accept that a gene duplication took place, followed by a translocation, we can understand that two duplicate α-chains were formed and evolved independently. One of them became the modern myoglobin while the other α-chain, had the property of dimerization, to form α_2 molecules. Ingram postulates that the genes of the α_2-chains duplicated again. After this

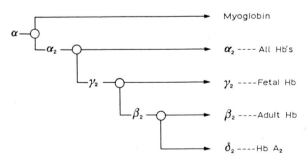

Fig. 7. Evolution of polypeptide chains in hemoglobins. The α-chain is the ancestral peptide chain; $\bigcirc$ indicates a point of gene duplication followed by translocation of the new gene. (Ingram, 1961)

duplication, two types of dimer—α_2 and γ_2—evolved sufficiently to form tetramers. The incidence of natural selection is, according to Ingram, to be found at the level of the heme–heme interaction, existing in the dimers and even more in the tetramers. This stage of hemoglobin evolution ($\alpha_2\gamma_2$) seems to have been reached in some bony fishes already possessing a four-chain hemoglobin. The next step in the hemoglobin phylogeny, in Ingram's scheme, was the duplication and translocation of the γ-chain gene. The new gene could develop along its own lines, providing a tetramer adapted to the physiological needs of the adult organism ($\alpha_2^A\beta_2^A$), while the old γ-chain continued to develop and provided a hemoglobin adapted to the oxygen exchanges of the fetus ($\alpha_2^A\gamma_2^A$). At this point three independent genes (α,β,γ) are assumed to have been present, each one capable of forming chains which dimerized and then aggregated to the tetramers $\alpha_2^A\beta_2^A$ or $\alpha_2^A\gamma_2^A$.

Hemoglobin A_2 is found in the blood of man and in that of a number of New World primates. Ingram and Stretton (1961) have suggested that comparatively recently the β gene underwent duplication and that the two resulting genes gave the present β and δ genes. Hemoglobin A_2 appears in this scheme as a new hemoglobin. The origin of the δ-chain is placed near the end of the evolutionary scheme, because of its great similarity to the β-chain. There is also genetic evidence that the genes for β- and δ-chains are linked.

In this scheme it is assumed that the different genes of human hemoglobin are all phylogenically derived from the gene of the ancestral α-chain and that the different peptide chains of human hemoglobin are phylogenically related. This assumption leads to the conclusion that the homology between primary structures can be recognized even in the case of a less pronounced degree of isology than was considered necessary in our previous discussion.

From the differences in primary structures of a specific chain in two vertebrates and from our paleontological knowledge, we may calculate the time elapsed since the differentiation of homologous chains. It may be estimated that the common ancestor of man and horse lived between 100 and 160 million years ago. The eighteen points of difference in the α-chain of man and horse led Zuckerkandl and Pauling (1962) to calculate a mean figure of 14.5 million years per amino acid substitution in a chain of about 150 amino acids.

In the scheme of the evolution of the amino acid chains by division and translocation of genes, the case of the lamprey, with its small hemoglobin similar to myoglobin, is considered to show that the cyclostomes diverged from the vertebrate phylogenic line before the first gene division in Ingram's scheme. It is interesting to compare these views on hemoglobin phylogeny, based on the modern techniques of determining amino acid sequences, with earlier ideas about "evolutionary trends" in hemoglobin, as formulated, for instance, in my book *L'Évolution Biochimique* (1944). Several hypothetical "evolutionary trends" were proposed there: "A carrier having the sigmoid type of curve can, therefore, be considered more highly evolved than a carrier conforming to the hyperbolic curve" (*Biochemical Evolution*, 1949, p. 28). "Haemoglobin of *Urechis* appears to be of a primitive type because it does not manifest the Bohr effect" (p. 31). "Since all cells of aerobic organisms have the ability to synthetize hematin catalysts (catalase, peroxidase, cytochrome system), they are potential producers of haemoglobin" (p. 29). Similar ideas were again expressed in a later review (Florkin, 1948), where another "trend" was also proposed. "L'ensemble des données acquises est en faveur de la notion selon laquelle les hémoglobines à affinité faible pour l'oxygène sont des transporteurs plus évolués que les hémoglobines à affinité forte" (Florkin, 1948, p. 183).

These inductions about evolutionary tendencies (hyperbolic curve more primitive than sigmoid curves, high oxygen affinity more primitive than low oxygen affinity and hemoglobin with no heme–heme interaction more primitive than those with Bohr effect) have been reported by Wald (1952) in a general review, and they are now sometimes designated as "Wald's theories" (Manwell, 1963) and attacked as such on the basis that these "primitive" or "specialized" aspects are "potentially" genetically "labile". They were certainly never claimed by myself to offer guide lines in phylogeny; a "primitive character" can very well result from the evolution of a "specialized character".

The belief is still general that the primitive hemoglobin was probably of relatively small size, and had therefore no heme–heme interaction, and thus that the "primitive" hemoglobins probably had hyperbolic dissociation curves.

To these evolutionary tendencies I have never myself ascribed any phylogenic meaning and I certainly should not decide that because there is no heme–heme interrelation in the botfly larva, *Arenicola* descends from it because its hemoglobin shows a Bohr effect. It occurs to nobody to claim that the parasitic worms are the ancestors of the free-living ones because they have no digestive enzymes.

In our study of the phylogeny of proteins, our only safe guide is the primary structure and in this light, it clearly appears, for instance, that the polypeptide chains of horse-heart cytochrome c and mammalian hemoglobin chains are not homologous. It is of course the homologies revealed by the determination of primary structures which should occupy us. A study of the hemoglobins of primates has provided very interesting results in this respect (Hill *et al.*, 1963). The degree of similarity accords with the more or less marked degree of antiquity of the species. The hemoglobins of the large monkeys differ very little from the hcmoglobin of man. The hemoglobins of the monkeys of Africa and of South America are also similar to human hemoglobin, though they show a relatively greater number of differences. Divergence from the human form is greatest in the most primitive primates, reaching its maximum in *Tupaia glis*. Hill and his collaborators have observed that, in comparison with man, the order of primates consistently displays greater variations in β- than in α-chain. Table III shows a striking similarity between the β-chain of primitive primates and the γ-chain of man. Hill concludes that the α-chains are more stable than the β-chains along the phyletic series of primates. This conclusion is in harmony with Ingram's scheme (Fig. 7) in which the phylogeny of the chains of hemoglobin is considered to have taken the following course: primitive chain— α-chain—γ-chain—β-chain—δ-chain.

Zuckerkandl and Pauling (1962) have suggested that the common molecular ancestor of α- and γ-chains must have appeared 600×10^6 years ago, the common ancestor of γ and β, 260×10^6 years ago and the ancestor of β and δ, about 44×10^6 years ago. Zuckerkandl

TABLE III

THE AMINO ACID COMPOSITION OF SOME PRIMATE TRYPTIC PEPTIDES FROM β-CHAINS[a] (Hill *et al.*, 1963)

	1	5	(βTp-I)	18	20	25	(βTp-III)
Human β-chain[b]	Val–his–leu–thr–pro–	glu–glu–lys	... val–asn–	val–asp–glu–	val–gly–gly–	glu–ala–leu–gly–arg–	
Hylobates lar	(Val, his, leu, thr, pro,	glu, glu, lys)...	(val, asp,	val, asp, glu,	val, gly, gly,	glu, ala, leu, gly, arg)	
Papio doguera			...(val, asp,	val, asp, glu,	val, gly, gly,	glu, ala, leu, gly, arg)	
Galago crassicaudatus	(Val, his, *phe*, thr, pro,	*gly, asp*, lys)...	(val, asp,	val, *glu*, glu,	val, gly, gly,	glu, ala, leu, gly, arg)	
Perodicticus potto	(Val, his, leu, thr, *glu, gly, asp*, lys)...		(val, asp,	val, asp, glu,	val, gly, gly,	glu, a'a, leu, gly, arg)	
Propithecus verreauxi			...(val, asp,	val, *glu, asp*,	*ala*, gly, gly,	glu, *thr*, leu, gly, arg)	
Lemur variegatus			...(val, asp,	val, *glu, lys*)			
L. catta			...(val, asp,	val, *glu, lys*,	val, gly, gly,	glu, *thr*, leu, gly, arg)	
L. fulvus			...(val, asp,	val, *glu, lys*,	val, gly, gly,	glu, ala, leu, gly, arg)	
			(γTp-I)				(γTp-III)
Human γ-chain[c]	Val–his– *phe*–thr–*glu*–	*glu–asp*–lys	... val–asn–	val–*glu–asp*–	*ala*–gly–gly–	glu–*thr*–leu–gly–arg–	

	31	35	(βTpIV)	45	50
Human β-chain	leu–leu–val–val–tyr–	pro–try–thr–glu–arg–		phe–phe–glu–ser–phe–gly–asp–leu–ser–thr–pro–	
Hylobates lar	(leu, leu, val, val, tyr,	pro, try, thr, glu, arg)		(phe, phe, glu, ser, phe, gly, asp, leu, ser, thr, pro,	
Papio doguera	(leu, leu, val, val, tyr,	pro, try, thr, glu, arg)		(phe, phe, glu, ser, *leu*, gly, asp, leu, *glu*, thr, pro,	
Galago crassicaudatus	(leu, leu, val, val, tyr,	pro, try, thr, glu, arg)		(phe, phe, glu, ser, *leu*, gly, asp, leu, ser, thr, pro,	
Perodicticus potto	(leu, leu, val, val, tyr,	pro, try, thr, glu, arg)			
Propithecus verreauxi	(leu, leu, val, val, tyr,	pro, try, thr, glu, arg)		(phe, phe, glu, ser, phe, gly, asp, leu, ser, *ser*, pro,	
Lemur variegatus	(leu, leu, val, val, tyr,	pro, try, thr, glu, arg)		(phe, phe, glu, ser, phe, gly, asp, leu, ser, *ser*, pro,	
L. catta	(leu, leu, val, val, tyr,	pro, try, thr, glu, arg)			
L. fulvus	leu, leu, val, val, tyr,	pro, try, thr, glu, arg			
			(γTp-IV)		
Human γ-chain	leu–leu–val–val–tyr–	pro–try–thr–glu–arg–		phe–phe–*asp*–ser–phe–gly–*asn*–leu–ser–*ser–ala*–	

TABLE III *(Continued)*

	55	(βTp-V)	(βTp-VI)	(βTp-VII)	70
Human β-chain	asp-ala-val-met-gly-asn-pro-lys-	val-lys-	ala-his-gly-lys-lys-val-leu-gly-ala-phe-ser-		
Hylobates lar	asp, ala, val, met, gly,asp, pro, lys)	(val, lys)	(ala, his, gly, lys)	(val, leu, gly, ala, phe, ser,	
Papio doguera	asp, ala, val, met, gly, asp, pro, lys)			(val, leu, glu, *ser*, phe, ser,	
Galago crassicaudatus	*glu*, ala, val, met, gly, asp, pro, lys)	(val, lys)	(ala,his, gly, lys)	(val, leu, *thr, ser,* phe, *gly,*	
Perodicticus potto		(val, lys)		(val, leu, *thr, ser,* phe, ser,	
Propithecus verreauxi	*ser, glu, ileu,* met, gly, asp, pro, lys)	(val, lys)		(val, leu, gly, *ser,* phe, ser,	
Lemur variegatus	*ser, ala, ileu,* met, gly, asp, pro, lys)	(val,lys)	(ala, his, gly, lys)	(val, leu, *thr, ser,* phe, *gly,*	
L. catta			(val,lys) (ala, his, gly, lys)		
L. fulvus			(val,lys)		

	(γTp-V)	(γTp-VI)	(γTp-VII)
Human γ-chain	*ser*–ala–*ileu*–met–gly–asp–pro–lys–val–lys–ala–his–gly–lys–		val–leu–*thr*–ser–*leu*–gly–

	75	80	(βTp-IX)
Human β-chain	asp-gly-leu-ala-his-leu-asp- asp-leu-lys		
Hylobates lar			
Papio doguera	asp, *glu,* leu, ala, his, leu, asp, asp, leu, lys)		
Galago crassicaudatus	asp, *asp,* leu, ala, his, leu, asp, asp, leu, lys)		
Perodicticus potto	asp, *ala, val,* ala, his, leu, asp, asp, leu, lys)		
Propithecus verreauxi	asp, *ala,* leu, *glu,* his, leu, asp, asp, leu, lys)		
Lemur variegatus	glu, *glu, thr, pro,* his, leu, asp, asp, leu, lys)		
L. catta			
L. fulvus			

	(γTp-IX)	(γTp-X)
Human γ-chain	asp–*ala–ileu–lys*–his–leu–asp–asp–leu–lys	

[a] The residues in italics are those which differ from analogous residues in human β-chains.
[b] Sequence according to Braunitzer et al. (1961).
[c] Sequence according to Schroeder et al. (1962).

(1965) has even been able to establish phylogenic trees of particular sites in the primary sequence of globin.

Braunitzer and Braun (1965) have isolated from the blood of the bloodworm *Chironomus thummi* four different hemoglobins. Three of these have been shown to be dimers of two different subnits (α- and β-chains). As the monomeric hemoglobin of *Lampetra fluviatilis* is made up of approximately 155 amino acids and the chains of *Chironomus* hemoglobin of about 130 residues, the authors suggest that in the phylogeny of hemoglobin a shortening of the amino acid chains has probably taken place. A similar conclusion can be drawn from the comparison of animal cytochrome *c* and of yeast cytochrome *c* (Fig. 1).

The comparative study of the amino acid sequences in fibrinopeptides, as accomplished by Doolittle and Blombäck (1964), is also of interest in the frame of phyletic studies. The gelification of mammalian blood plasma in the course of blood clotting is the result of the action of an enzyme, thrombin, formed during the coagulation process, on fibrinogen, a protein present in the circulating plasma. Thrombin generally splits off two pairs of peptides from fibrinogen (fibrinopeptides A and B) and this splitting frees the groups which

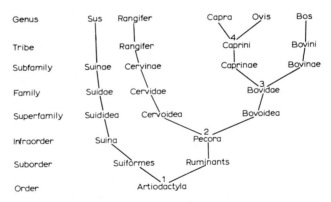

Fig. 8. Phylogenic relationship in the five artiodactyls, the fibrinopeptide structures of which have been elucidated. *Sus*, pig; *Rangifer*, reindeer; *Capra*, goat; *Ovis*, sheep; and *Bos*, ox. Approximate chronological junctions are indicated by numbers: 1, Eocene; 2, Oligocene; 3, Miocene; 4, Pliocene. (Doolittle and Blombäck, 1964)

	19	18	17	16	15	14	13	12	11	10	9	8	7	6	5	4	3	2	1
Ox	H – Glu – Asp – Asp – Ser – Asp – Pro – Pro – Ser – Gly – Asp – Phe – Leu – Thr – Glu – Gly – Gly – Gly – Val – Arg – OH
Sheep	H – Ala – Asp – Gly – Ser – Asp – Pro – Val – Gly – Gly – Glu – Phe – Leu – Ala – Glu – Gly – Gly – Gly – Val – Arg – OH
Goat	H – Ala – Asp – Asp – Ser – Asp – Pro – Val – Gly – Gly – Glu – Phe – Leu – Ala – Glu – Gly – Gly – Gly – Val – Arg – OH
Reindeer	H – Ala – Asp – Gly – Ser – Asp – Pro – Ala – Gly – Gly – Glu – Phe – (Leu, Ala, Glu, Gly, Gly, Gly, Val) Arg – OH
Pig	H – Ala – Glu – Val – Gln – Asp – Lys – Gly – Glu – Phe – Leu – Ala – Glu – Gly – Gly – Gly – Val – Arg – OH
Human	H – Ala – Asp – Ser – Gly – Glu – Gly – Asp – Phe – Leu – Ala – Glu – Gly – Gly – Gly – Val – Arg – OH
Rabbit	H – Val – Asp – Pro – Gly – Glu – Thr – Ser – Phe – Leu (Thr, Glu, Gly, Gly) Asp – Ala – Arg – OH

Fig. 9. Proposed amino acid sequences of fibrinopeptides A from seven mammals. (Doolittle and Blombäck, 1964)

21 20 19 18 17 16 15 14 13 12 11 10 9 8 7 6 5 4 3 2 1

Ox

$$\overset{\text{SO}_4}{|}$$

Pyr–Phe–Pro–Thr–Asp–Tyr–Asp–Glu–Gly–Gln–Asp–Asp–Arg–Pro–Lys–Val–Gly–Leu–Gly–Ala–Arg–OH

Sheep

$$\overset{\text{SO}_4}{|}$$

Gly–Tyr–Leu–Asp–Tyr–Asp–Glu–Val–Asp–Asp–Asn–Arg–Ala–Lys–Leu–Pro–Leu–Asp–Ala–Arg–OH

Goat

$$\overset{\text{SO}_4}{|}$$

Gly–Tyr–Leu–Asp–Tyr–Asp–Glu–Val–Asp–Asp–Asn–Arg–Ala–Lys–Leu–Pro–Leu–Asp–Ala–Arg–OH

Reindeer

$$\overset{\text{SO}_4}{|}$$

Pyr–Leu–Ala–Asp–Tyr–Asp–Glu–Val (Glu, His, Asp) Arg–Ala–Lys–Leu–His–Leu–Asp–Ala–Arg–OH

Pig

Ala–Ile–Asp–Tyr–Asp–Glu–Asp–Glu–Asp–Gly–Arg–Pro–Lys–Val–His–Val–Asp–Ala–Arg–OH

Human

$$\overset{\text{SO}_4}{|}$$

Pyr–Gly–Val–Asn–Asp–Asn–Glu–Glu–Gly–Phe–Phe–Ser–Ala–Arg–OH

Rabbit

Ala–Asp–Asp–Tyr (Asp, Glu, Pro, Leu, Asp, Val) Asp–Ala–Arg–OH

Fig. 10. Proposed amino acid sequences of fibrinopeptides B from seven mammals. Pyr = pyrrolidone carboxylic acid. (Doolittle and Blombäck, 1964)

associate the fibrinogen molecules into fibrin by means of hydrogen bonds. Doolittle and Blombäck have studied the amino acid sequences in a series of artiodactyls, the phyletic relations of which are represented in Fig. 8. Figs. 9 and 10 show the sequences of amino acids in fibrinopeptides A and B of a number of these mammals. The two species most closely related in this phylogenic tree are the goat (*Capra*) and the sheep (*Ovis*), which both exhibit identical sequences in the two fibrinopeptides. When the degrees of correspondence of these sequences of amino acids are plotted against the time elapsed since the animals concerned had a common ancestor (according to commonly accepted phylogeny), a smooth curve is obtained, except in the case of the relation of goat and sheep to *Rangifer* and ox.

At the C-terminal end of the fibrinopeptide B chain, the five species show a similar structure. In position 3 (Fig. 11) they all have an aspartate residue with the exception of the ox, which has a residue of glycine in this position. It may be postulated that aspartic acid has been replaced by glycine after the separation of the bovine branch.

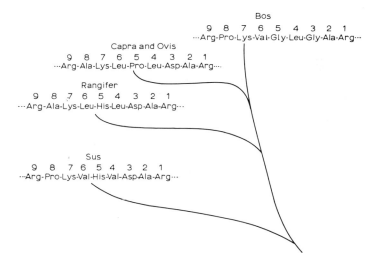

Fig. 11. Stepwise amino acid substitution in one region of fibrinopeptide B in five artiodactyls. (Doolittle and Blombäck, 1964)

Bibliography p. 32

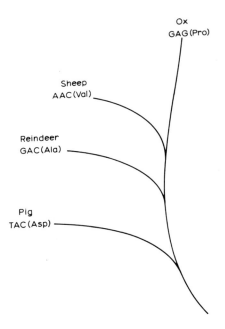

Fig. 12. Possible mode of base substitution in DNA which gave rise to amino acid changes at position A-13 of artiodactyl fibrinopeptide A. A = adenine; C = cytosine; G = guanosine; T = thymine. (Doolittle and Blombäck, 1964)

On the other hand, all the peptides concerned, with the exception of the peptide of the pig, show in position 4 a leucine residue. It is therefore under the function of ruminants and non-ruminants that the replacement of leucine has been accomplished. In Fig. 12 the changes in position A-13 during the 40 million years of the evolution of artiodactyls since the Eocene have been considered and the corresponding sequences of the genetic code (Eck, 1963) have been added. Jukes (1965) has also reported on the changes, during the evolution of the hemoglobin genes, in the composition of the triplets involved in the amino acid changes at the level of the hemoglobins.

If it is accepted that a generation of artiodactyls has an average life-span of approximately five years, it may be concluded that, with respect to the combination of purine and pyrimidine bases corresponding to the amino acid in position A-13, a persisting

TABLE IV

SURVEY OF A NUMBER OF DIFFERENCES IN THE AMINO ACID SEQUENCES OF
CYTOCHROME c IN VARIOUS SPECIES RELATED TO THE TIME ELAPSED SINCE
THEIR DIFFERENTIATION FROM A COMMON ANCESTOR

(Smith and Margoliash, 1964)

Species	Number of differences	Time elapsed (years)
Man—Monkey	1	$50–60 \times 10^6$
Man—Horse	12	$70–75 \times 10^6$
Man—Dog	10	$70–75 \times 10^6$
Pig—Cow—Sheep	0	
Horse—Cow	3	$60–65 \times 10^6$
Mammal—Chicken	10–15	28×10^7
Mammal—Tuna fish	17–21	40×10^7
Vertebrate—Yeast	43–48	$1–2 \times 10^9$

substitution of base has been accomplished once in the course of 3×10^6 generations.

Blombäck *et al.* (1965) have extended these comparative studies to a larger number of mammals and have concluded that the B peptides are more rapidly modified in evolution than are A peptides.

With respect to cytochrome c, Smith and Margoliash (1964) have compiled a table (Table IV), in which the number of persisting differences in the amino acid sequence is related to the length of time elapsed since the separation of the species considered.

On the other hand it must be emphasized that the cytochrome c, of vertebrates and yeast coincides in 58 loci—a circumstance which cannot be due to chance, *i.e.* to the phenomenon of convergence (see Fig. 1).

BIBLIOGRAPHY

ACHER, R., The comparative chemistry of neurohypophyseal hormones, in: H. HELLER (Ed.), *Comparative Aspects of Neurohypophyseal Morphology and Function*, Symp. Zool. Soc. London, 9 (1963) 83–106.

ACHER, R., J. CHAUVET, M. T. CHAUVET AND D. CREPY, Caractérisation des hormones neurohypophysaires d'un poisson osseux d'eau douce, la carpe (*Cyprinus carpio*). Comparaison avec les hormones des poissons osseux marins, *Comp. Biochem. Physiol.*, 14 (1965a) 245–254.

ACHER, R., J. CHAUVET, M. T. CHAUVET AND D. CREPY, Phylogénie des peptides neurohypophysaires: Isolement d'une nouvelle hormone, la glumitocine (Ser_4 –Gln_8–ocytocine) présente chez un poisson cartilagineux, la raie (*Raia clavata*). *Biochim. Biophys. Acta*, 107 (1965b) 393–396.

ANFINSEN, C. B., *The Molecular Basis of Evolution*, Wiley, New York, 1959.

BLOMBÄCK, B., R. F. DOOLITTLE AND M. BLOMBÄCK, Fibrinogen: structure and evolution, in: H. PEETERS (Ed.), *Protides of the Biological Fluids*, Vol. 12, Elsevier, Amsterdam, 1965, pp. 87–94.

BRAUNITZER, G., AND V. BRAUN, Zur Phylogenie des Hämoglobinmoleküls. Untersuchungen an Insekten-Hämoglobinen (*Chironomus thummi*), *Z. Physiol. Chem.*, 340 (1965) 88–91.

BRAUNITZER, G., R. GEHRING-MÜLLER, N. HILSCHMANN, K. HILSE, G. HOBOM, V. RUDLOFF AND B. WITTMANN-LEIBOLD, Die Konstitution des normalen adulten Humanhämoglobins, *Z. Physiol. Chem.*, 325 (1961) 283–286.

BURGERS, A. C. J., Occurrence of 3 electrophoretic components with melanocyte-stimulating activity in extracts of single pituitary glands from ungulates, *Endocrinology*, 68 (1961) 698–703.

DOOLITTLE, R. F., AND B. BLOMBÄCK, Amino acid sequence investigations of fibrinopeptides from various mammals: evolutionary implications, *Nature*, 202 (1964) 147–152.

ECK, R. V., Genetic code: emergence of a symmetrical pattern, *Science*, 140 (1963) 477–481.

FLORKIN, M., *L'Évolution Biochimique*, Masson, Paris, 1944.

FLORKIN, M., La biologie des hématinoprotéides oxygénables, *Experientia*, 4 (1948) 176–191.

FLORKIN, M., *Biochemical Evolution* (translated by S. MORGULIS), Academic Press, New York, 1949.

FLORKIN, M., Isologie, homologie, analogie et convergence en biochimie comparée, *Bull. Classe Sci. Acad. Roy. Belg.*, 48 (1962) 819–824.

FLORKIN, M., Quelques perspectives nouvelles de la biochimie comparée, *Bull. Soc. Chim. Biol.*, 45 (1963a) 653–680.

FLORKIN, M., L'évolution biochimique et la radiation physiologique des systèmes biochimiques chez les animaux, in: A. I. OPARIN (Ed.), *Evolutionary Biochemistry*, Pergamon, Oxford, 1963b, pp. 250–270.

FLORKIN, M., Perspectives in comparative biochemistry, in: C. A. LEONE (Ed.), *Taxonomic Biochemistry and Serology*, Ronald, New York, 1964, pp. 51–74.

FLORKIN, M., On the phylogeny of proteins, in: H. PEETERS (Ed.), *Protides of the Biological Fluids*, Vol. 12, Elsevier, Amsterdam, 1965, pp. 17–28.

FOX, H. M., On chlorocruorin and haemoglobin, *Proc. Roy. Soc. (London), Ser. B*, 136 (1949–50) 378–388.

HARRIS, I., Chemistry of pituitary polypeptide hormones, *Brit. Med. Bull.*, 16 (1960) 183–188.

HILL, R. L., J. BUETTNER-JANUSCH AND V. BUETTNER-JANUSCH, Evolution of hemoglobin in primates, *Proc. Natl. Acad. Sci.*, *U.S.*, 50 (1963) 885–893.

INGRAM, V. M., Gene evolution and the haemoglobins, *Nature*, 189 (1961) 704–708.

INGRAM, V. M., AND A. O. W. STRETTON, Human haemoglobin A$_2$: chemistry, genetics and evolution, *Nature*, 190 (1961) 1079–1084.

JUKES, T. H., Coding triplets in the evolution of hemoglobin and cytochromes *c* genes, in: S. W. FOX (Ed.), *The Origins of Prebiological Systems*, Academic Press, New York, 1965, pp. 407–435.

MANWELL, C., The blood proteins of cyclostomes. A study in phylogenetic and ontogenetic biochemistry, in: A. BRODAL AND R. FÄNGE (Eds.), *The Biology of Myxine*, Universitetsforlag, Oslo, 1963, pp. 372–455.

SCHROEDER, W. A., J. R. SHELTON, J. B. SHELTON AND J. CORMICK, Further sequences in the γ-chain of human fetal hemoglobin, *Proc. Natl. Acad. Sci.*, *U.S.*, 48 (1962) 284–287.

SMITH, E. L., AND E. MARGOLIASH, Evolution of cytochrome *c*, *Federation Proc.*, 23 (1964) 1243–1247.

SMYTH, D. G., W. H. STEIN, AND S. MOORE, The sequence of amino acid residues in bovine pancreatic ribonuclease: revisions and confirmations, *J. Biol. Chem.*, 238 (1963) 227–234.

WALD, G., Biochemical evolution, in: E. S. G. BARRON (Ed.), *Modern Trends in Physiology and Biochemistry*, Academic Press, New York, 1952, pp. 337–376.

ZUCKERKANDL, E., Further principles of chemical paleogenetics as applied to the evolution of hemoglobin, in: H. PEETERS (Ed.), *Protides of the Biological Fluids*, Vol. 12, Elsevier, Amsterdam, 1965, pp. 102–109.

ZUCKERKANDL, E., AND L. PAULING, Molecular disease, evolution and genic heterogeneity, in: M. KASHA AND B. PULLMAN (Eds.), *Horizons in Biochemistry*, Academic Press, New York, 1962, pp. 189–225.

Chapter 4

Biosynthesis and Phylogeny

The metabolism of cells comprises an anabolic phase and a catabolic phase. The molecules of carbohydrates go through the metabolic sequence of glycolysis and reach the stage of pyruvate. In anaerobiosis, pyruvate gives ethyl alcohol or lactic acid, while in the presence of oxygen it leads to acetyl-CoA (C_2 unit) and is, through the stages of the Krebs cycle, transformed into carbonic anhydride and water. The fatty acids, through Lynen's cycle, are also catabolized with the formation of acetyl-CoA (Fig. 13). With respect to amino acids, the cetonic acids resulting from their deamination are introduced at several points of the catabolic path described above, in the course of which are synthetized molecules of ATP in which the energy is stored for several forms of work, including biosynthesis.

This biosynthesis is for a large part kept going by a pool of units of small size resulting from the catabolism of carbohydrates, fatty acids and cetonic acids derived from amino acids, whether these sources are of exogenous or endogenous origin.

By the use of stable and radioactive isotopes, of auxotrophic mutants of microorganisms, and of purified enzymes, the metabolic pathways of a number of biosyntheses have been understood. One

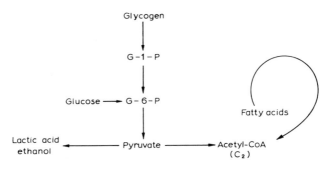

Fig. 13. Glycolysis and two of the sources of C_2 units.

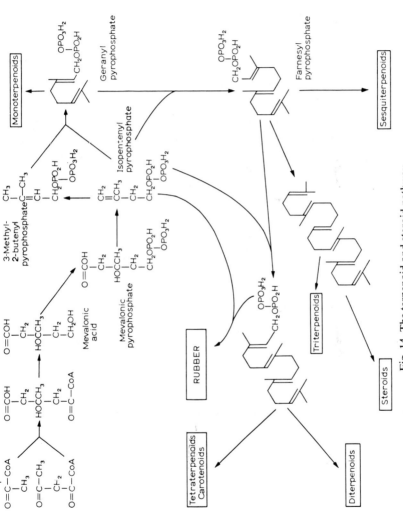

Fig. 14. The terpenoid and steroid pathway.

of the paths starts with C_2 units (acetyl-CoA) and can be called the terpenoid and steroid pathway (Fig. 14). This path leads to mevalonic acid, the starting point of the biosynthesis of the different terpenes which are constituents of the essential oils of plants. These constituents play an important role in the relationships between plants and insects. They are compounds composed of isopentane units and contain 5, 10, 15, 20 or more carbon atoms and are called respectively hemiterpenes, mono-, sesqui-, and di- or polyterpenes. From the material which cannot be distilled in steam, it is possible to obtain by solvent extraction a series of other substances containing 20, 30, 40 or more carbon atoms, and belonging to the group of diterpenes (*e.g.* the resins), triterpenes (*e.g.* the saponins), tetraterpenes (*e.g.* the carotenoids) or to the polyterpenes (*e.g.* rubber). Moreover, a whole series of organic compounds synthetized by plants are related to isopentane, of which they contain varying numbers of units in their structure. Among the isoprenoids are the irones. Many monoterpenes are found in plants and, in general, but not always, one can consider their formula as being based on two isopentane units joined in head and tail union. The cyclic monoterpenes and sesquiterpenes can be considered to result from the rolling up of the same chains. Certain diterpenes may be regarded as containing four isopentane units in head-to-tail union. This is the case with phytol and vitamin A; others have an irregular arrangement. Regarding tetraterpenes, plants are able to synthetize carotenoid molecules, while animals are only able to modify them, for example, by oxidation. Astaxanthin, a carotenoid usually found in crustaceans, is one such oxidation product.

In mammals, birds and certain amphibians, the ingestion of carotenoids in the food results in the absorption of carotene in the intestine, the extent of absorption depending on the greater or lesser activity of the intestinal carotenase, which converts carotene to vitamin A. As a result, the reserve fats become more or less saturated with carotene. Man and other primates absorb carotenoids in general, as does the frog; other animals exercise a selective absorption. For example, the horse and the cow selectively absorb carotenoids and store them without alteration; birds and fish show a preference for xanthophylls. However, birds and fish modify one of the ingested xanthophylls, lutein, and the products of the oxidation are deposited in the feathers in the case of birds, and in the skin in the case of fish. The carotenoid structure appears to be connected in a general way with the function of photoreception. The most primitive type of photoreception, lacking any differentiated photoreceptors, is the type called dermatoptic, which is found in primitive types of organisms, up to the amphibians, and also in plants. The maximum sensitivity of this dermatoptic function is in the ultraviolet part of the spectrum, around 365 mμ; it is operative in photokinetic processes involving tropisms towards light in the above-named primitive types. Now, in a number of cases, photoreceptors have evolved secondarily and developed new kinds of receptor molecules adapted to the light from the sun and the sky. All these substances belong to the carotenoid group. In plants, phototropic bending depends on the properties of carotenoids such as xanthophyll in *Avena,* or β-carotene in the sporangiophores of *Phycomyces.* The orientation of an animal depends on visual photoreception and requires the presence of other carotenoids showing the same kind of adaptation to sunlight and having a maximum absorption at around 500 mμ. This development is due to the ability of plants, mentioned above, to change some of the carotenoids into vitamin A. There are two types of vitamin A: vitamin A_1 and vitamin A_2.

Let us briefly consider the general system of photoreception in the eye of land vertebrates including birds. The pigment of their retina is rhodopsin, a rose-colored carotenoid–protein complex. In aqueous solution, its absorption spectrum consists of a single broad band with a maximum at 500 mμ. In light, it is bleached to orange and yellow pigments, and in the process the carotenoid retinene I is liberated. The latter substance has never been found anywhere except in the retina. Its spectrum in chloroform consists of a single band with a maximum at about 387 mμ. In the retina, the mixture of retinene I and protein reverts to rhodopsin and, in addition, retinene I is converted to vitamin A_1, which in the intact eye also reunites with protein to form rhodopsin. This system is

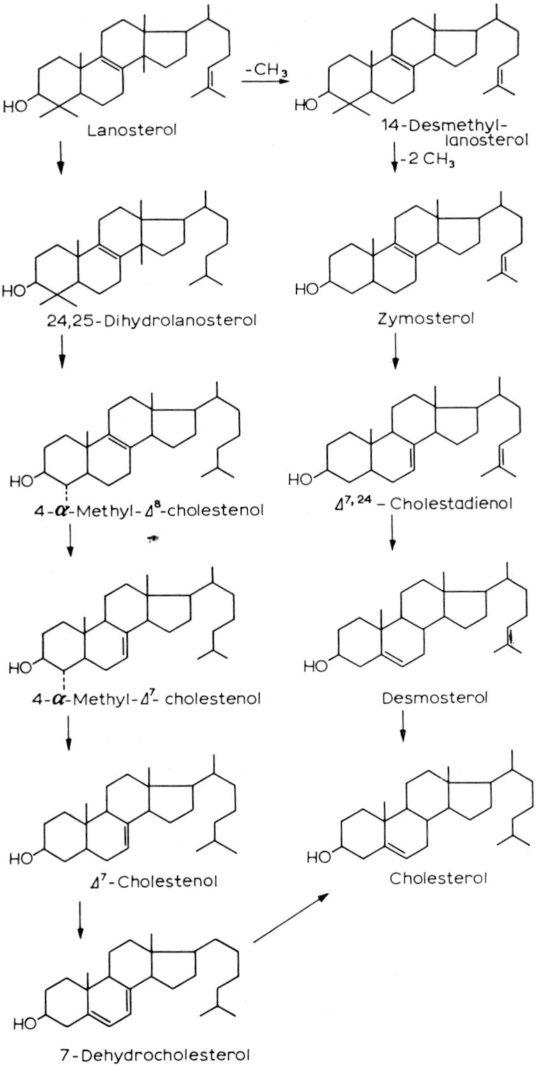

Fig. 15. Pathways of cholesterol biosynthesis. (Richards and Hendrickson, 1964)

not only present in the eye of mammals and birds but also in that of some invertebrates, such as the squid *Loligo* and the crayfish *Cambarus*.

If we consider the system in the eye of marine fish, we find the rhodopsin system as in birds and mammals and invertebrates, but this is not the system to be found in the eye of freshwater fish which utilize another system, the porphyropsin system. Porphyropsin, like rhodopsin, is a carotenoid–protein complex and is purple in color. Its spectrum resembles that of rhodopsin, but has a maximum at 522 mμ. On exposure to light, a substance having properties similar to rhodopsin is liberated; it is called retinene II. In chloroform it has an absorption maximum at 405 mμ. In the retina it is converted simultaneously to porphyropsin and to vitamin A_2.

From squalene lanosterol is derived, and from the latter two pathways lead to cholesterol (Fig. 15). Starting from steroids, specialized pathways lead, in vertebrates, to steroid hormones or to bile acids. The example of the terpenoid and steroid pathway (Fig. 14) shows how in biosynthetic pathways new syntheses may result from lateral or terminal extensions on a definite path, which becomes longer or more branched.

This concept could be developed when considering a number of pathways grafted on the common tree of cellular anabolisms starting from low molecular precursors. Each of these pathways with its ramifications show the diversity of variants embroidered by evolution on the common theme from which the branches originate.

The biosynthesis of porphyrins starts at the level of glycine and of succinyl-CoA derived from the Krebs cycle (Fig. 16). The δ-aminolevulinic acid and the porphobilinogen are situated on this pathway.

By insertion of iron into the protoporphyrin nucleus we obtain what Granick has called the "iron branch" of the biosynthetic chain. Protoporphyrin is also the starting point for the biosynthesis of chlorophyll (the "magnesium branch" of Granick) as well as for the biosynthesis of heme.

A cell capable of photosynthesis contains at least one chlorophyll and at least one yellow pigment. In addition it often contains

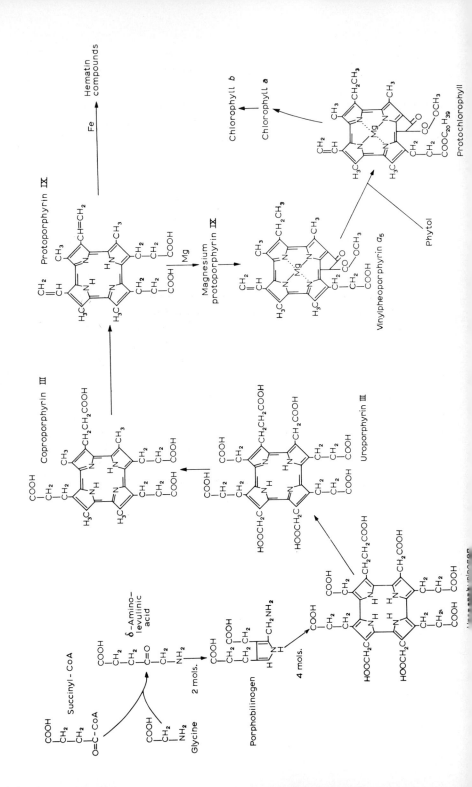

a phycobilin. The chief pigment in photosynthesis, both in algae and in the higher plants, is chlorophyll a. In the photosynthetic bacteria, on the other hand, we find a different chlorophyll, bacteriochlorophyll. Whereas in green plants the chloroplasts contain chlorophyll a and chlorophyll b, in the algae we find a number of combinations: $a + b, a + c, a + d, a + e$. In addition we sometimes find a phycobilin. The phycobilins, which are soluble in water, are proteins combined with a chromophore belonging to the class of bile pigments. The phycoerythrins are predominant in the red algae and the phycocyanins in the blue–green algae.

The chromophore of the phycoerythrins, phycoerythrobilin, is identical to mesobilierythrin; the chromophore of the phycocyanins is mesobiliviolin (Fig. 17). Various phycoerythrins are found in algae, which differ in the structure of the protein moiety; R-phycoerythrin is the most common and is found in the *Rhodophyceae*, whilst C-phycoerythrin is present in the *Myxophyceae*. Among the phycocyanins, R-phycocyanin is present in the *Rhodophyceae* and C-phycocyanin in the *Myxophyceae*. The phycobilins serve to absorb light and transmit energy to other systems, notably by the chlorophyll system. Phycoerythrins and phycobilins appear to result from a lateral chain of the porphyrin pathway starting at the level of hematin compounds (see Bogorad, 1963).

Plants are able to effect the synthesis of porphyrins along the "iron branch" and along the "magnesium branch", whilst in animals

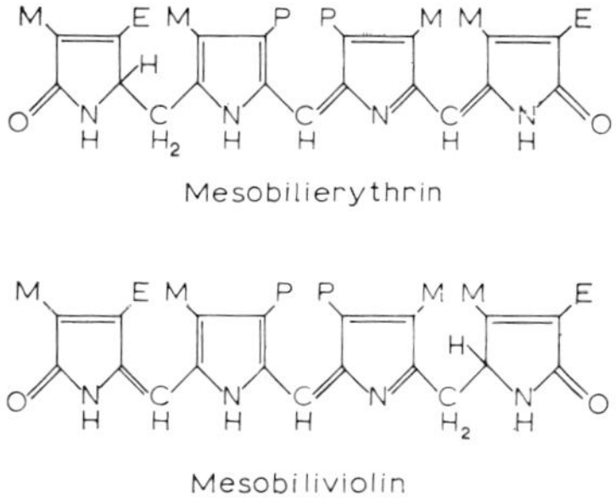

Mesobilierythrin

Mesobiliviolin

Fig. 17. Mesobilierythrin and mesobiliviolin (M = methyl group; E = ethyl group; P = propyl group).

Bibliography p. 49

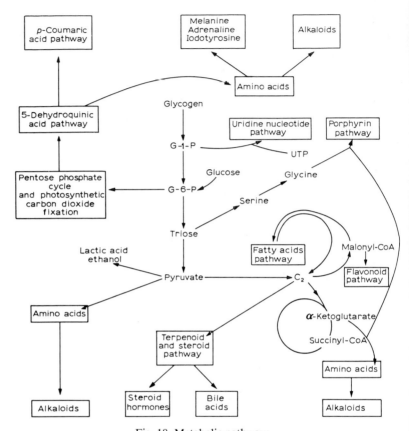

Fig. 18. Metabolic pathways.

the latter is lacking. However, animals have particularly developed the "iron branch" as far as the biosynthesis of the compound of heme and globin, hemoglobin, is concerned. The biosynthesis of hemoglobin is sometimes observed in plants, for example in the root nodules of legumes. In animals, the presence of hemoglobin other than in the blood has often been demonstrated.

In the outer frames of Fig. 18, which gives a general picture of biosynthesis, are found the most specialized branches, which developed last. For instance, the flavonoid pathways (Fig. 19) are mostly found in angiosperms. The coumarins produced in the 5-

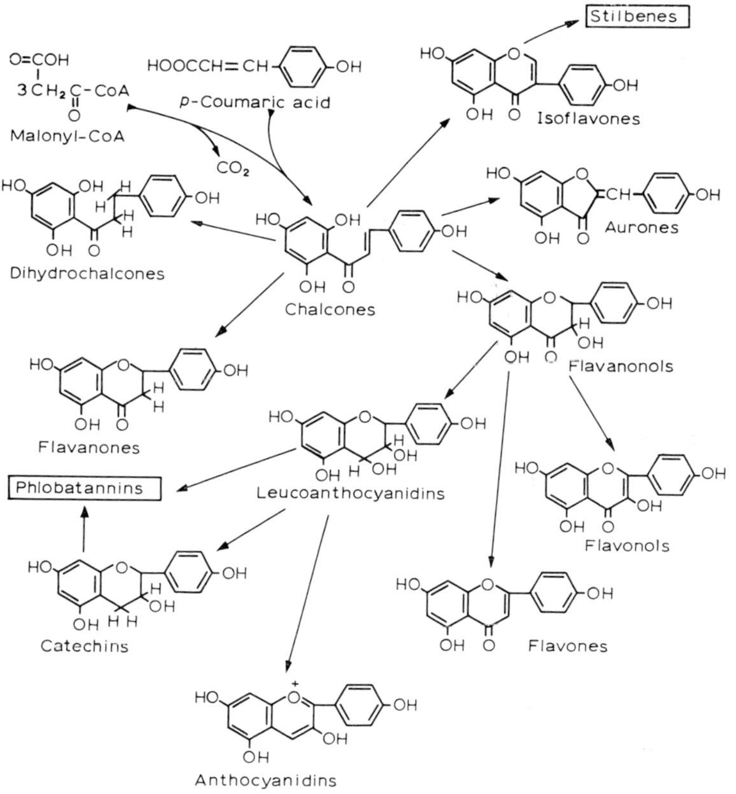

Fig. 19. Flavonoid pathways. (Robinson, 1963)

hydroquinic acid pathway (Fig. 20) are found in dicotyledons and many examples of this kind could be cited with respect to many of the branches of Fig. 18.

When we considered the concept of homology (Chapter 2), we defined it as pointing to a common origin, starting from an initial prototype, and we decided that a high degree of isology in the primary structure of proteins, *viz.* in amino acid sequences, would be taken as a sign of homology. Homology may be extended, as we have already said, to more or less identical chains of homologous enzymes and the final products of the biosynthetic action of these homolo-

Bibliography p. 49

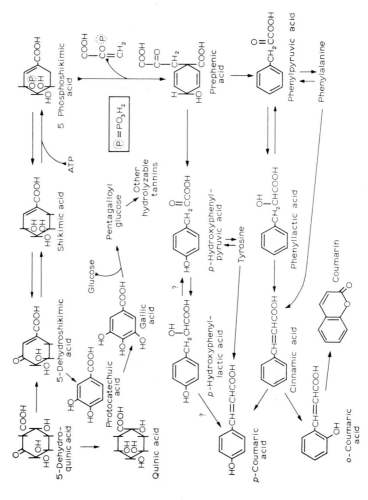

Fig. 20. 5-Dehydroquinic acid pathway. (Robinson, 1963)

gous enzyme chains may also be considered homologous. But while the homology attested by a high degree of isology in the primary structure of nucleic acids or amino acid sequences can be considered a homology of the first degree, or a direct homology, isologous substances biosynthetized by homologous enzyme chains are to be considered to exhibit an indirect, or second-degree, homology. What is homologous in the case of the cholesterol isolated from several organisms, is the chain of homologous enzymes catalyzing the biosynthesis.

Benzoic acid is not homologous in all its biological localizations. It may be derived from shikimic acid, but in other cases it is produced by a cyclization. Lysine, in certain microorganisms, is biosynthetized from a,a'-diaminopimelic acid, while in other microorganisms it derives from a-aminoadipic acid.

If it is true that the porphyrins derived from hemoglobins, cytochrome c, etc. in different organisms are in general derived from δ-

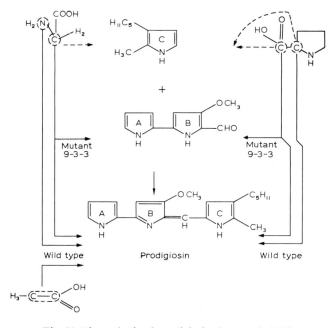

Fig. 21. Biosynthesis of prodigiosin. (Bogorad, 1963)

aminolevulinic acid and from porphobilinogen as shown in Fig. 16, prodigiosin is not biosynthetized through the same pathway (Fig. 21).

These are examples of biochemical convergence—a concept to be distinguished from the eventual presence of homologous enzyme chains and, consequently, of homologous biosynthetic products in different branches of phylogeny. This will be designated as parallelism—a consequence of the inheritance of certain genes by a large portion of the system of living organisms.

A diagram such as Fig. 18 shows that the organisms, in the course of their diversification and of the realization of their adaptations, have not indefinitely multiplied the number of biosynthetic pathways, many of which have grown according to the system of extensions on common biosynthetic lines. This aspect, which of course displays many exceptions, is an aspect of *biosynthetic economy* which has a definite impact on the phylogeny of many biosynthetic systems and of their products.

If we consider the terpenoid and steroid metabolic pathway, no longer from the viewpoint of the biochemical synthesis of natural products, but from that of phylogeny, *i.e.* guided by the principle we have already stated, we start from the phylogeny of the organisms

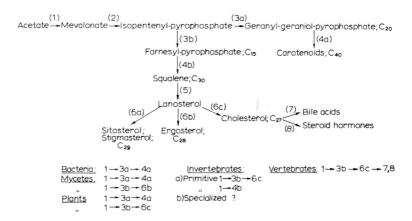

Fig. 22. Phylogeny of metabolic sequences in the mevalonic and steroid pathway. (Bloch, 1964)

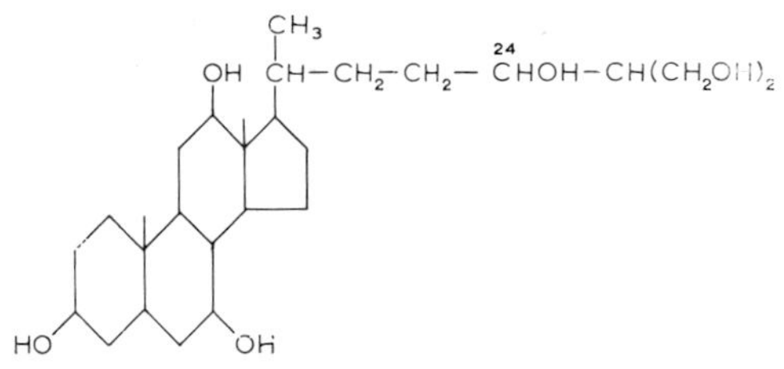

Fig. 23. Scymnol.

and follow, along the branches of the phylogenic trees, the possible changes at the level of molccular units. These changes, as we see in Fig. 22 (Bloch, 1964), show that the metabolic pathways are not identical in all groups of classification, but that in each case biosynthesis repeats the phylogeny of the molecule, *i.e.* the changes of structure it has undergone along the phylogenic branches. If we consider, for instance, in Fig. 22, the pathway leading to bile acids—a pathway which arises in vertebrates, we can see that cholesterol, itself resulting from the operation of a long-chain enzyme system, is probably converted into primary alcohol (A/B *cis*) containing supplementary hydroxyl groups. This alcohol is of the kind found conjugated with sulfate in the bile of selacians, teleosteans and amphibians. One of the alcohols of this type is scymnol which had been extracted from the bile of sharks (Fig. 23).

The primary alcoholic function carried by C-27 is thereafter oxidized to give hydroxylized coprostanic acids such as those found in amphibians and inferior reptiles. Finally, a β-oxidation of the lateral chain gives C_{24} bile acids, characterizing the bile of superior reptiles, birds and mammals.

In mammals, these bile acids are conjugated with glycine (see Haslewood, 1962). In the chain of biosynthesis of bile acids, it is clear that biosynthesis repeats phylogeny.

If we consider an organic molecule as it exists in different organisms, we can, in spite of the isology, only consider the molecule as homologous if the chain of enzymes catalyzing the biosynthesis is itself homologous, *i.e.* made up of an identical chain of homologous enzyme proteins. Complete information in this field will have to

Bibliography p. 49

await the collection of a large number of primary sequences. This could be greatly speeded up by the construction of automated laboratories, the services of which could be at the disposal of a large number of scientists.

If we consider, in glycolysis, the enzymatic step leading to two triose phosphates by scission of a molecule of fructose diphosphate, it may be shown that two types of aldolase exist and that they are probably not homologous according to the definition given above (Rutter, 1964). They are analogous molecules, probably acting through different mechanisms. In bacteria, yeasts, mycetes and blue–green algae, we find type II, while green algae, protozoa, plants and animals have type I. The two groups coexist in *Euglena* and in *Chlamydomonas*. It is very probable that aldolase II has been replaced by aldolase I at a point in phylogeny that can be defined. This variation does not prevent us from accepting that the enzyme chain of glycolysis forms a system the homology of which reigns among the generality of cells.

In Fig. 22 each enzyme chain leading to each metabolic product can be individualized. Chain 1 → 3b → 6c → 7 expresses the phylogeny of bile acids, while the chain 1 → 3b → 6c → 8 symbolizes the biosynthesis of steroid hormones. Of course, many new data on the phylogeny of proteins and on the biosynthetic chains will be necessary before it will be possible to determine to what extent the course of the pathways of the biosynthesis of molecules recapitulates the phylogeny of these molecules.

If, as is too often the practice in works on "plant chemotaxonomy", the phylogeny of organisms continues to be inferred from the distribution of organic constituents, the field will be progressively obscured. The phylogeny of proteins, as we have seen in Chapter 3, has to be considered at the level of the genes. We can also see that the biosynthetic chains are modified by the addition of new enzymes along the branches of the phylogenic tree of organisms. Modifications of this nature, in spite of their being encountered in adaptive radiations and in complex physiological relations and mechanisms, are, by the fact that they consist of the addition of a few protein catalysts, clearly under the influence of a small number of specific genes.

BIBLIOGRAPHY

BLOCH, K., Comparative aspects of lipid metabolism, in: C. A. LEONE (Ed.), *Taxonomic Biochemistry and Serology*, Ronald, New York, 1964, pp. 377–390.

BOGORAD, L., The biogenesis of heme, chlorophylls and bile pigments, in: P. BERNFELD (Ed.), *Biogenesis of Natural Compounds*, Pergamon, Oxford, 1963, pp. 183–231.

HASLEWOOD, G. A. D., Bile salts. Structure, distribution and possible biological significance as a species character, in: M. FLORKIN AND H. S. MASON (Eds.), *Comparative Biochemistry*, Vol. 3, Academic Press, New York, 1962, pp. 205–229.

RICHARDS, J. H., AND J. B. HENDRIKSON (Eds.), *The Biosynthesis of Steroids, Terpenes and Acetogenins*, Benjamin, New York, 1964, p. 322.

ROBINSON, T., *The Organic Constituents of Higher Plants*, Burgess, Minneapolis, 1963, p. 271.

RUTTER, W. J., Evolution of aldolase, *Federation Proc.*, 23 (1964) 1248–1257.

Chapter 5

Chitin, Chitinolysis and Phylogeny*

While the protective barrier of the body of vertebrates is made of keratin, a complex of chitin and proteins plays the same role in arthropoda. Chitin (Fig. 24) is a homopolysaccharide, *i.e.* a homoglycan of linear structure (Horton and Wolfrom, 1962). Mild acid hydrolysis of chitin leads to the diholoside of N-acetyl-D-glucosamine (Fig. 24) called chitobiose (Zechmeister and Toth, 1931) as

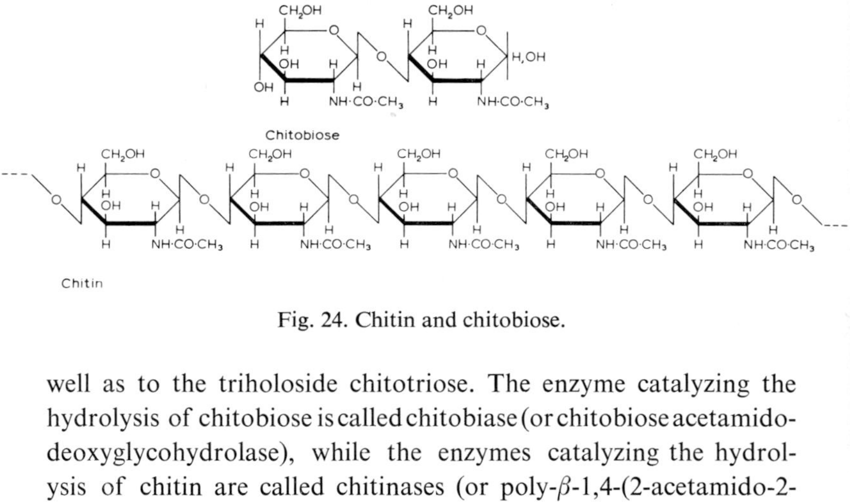

Chitobiose

Chitin

Fig. 24. Chitin and chitobiose.

well as to the triholoside chitotriose. The enzyme catalyzing the hydrolysis of chitobiose is called chitobiase (or chitobiose acetamidodeoxyglycohydrolase), while the enzymes catalyzing the hydrolysis of chitin are called chitinases (or poly-β-1,4-(2-acetamido-2-deoxy)-D-glucoside glycanohydrolases). The complete enzymatic hydrolysis of chitin leading to a complete transformation into acetylglucosamine can be represented as in Fig. 25.

Fig. 25.

---

* Chapter 5 has been revised by my colleague Dr. Ch. Jeuniaux.

As indicated in Fig. 25, a small amount of free acetylglucosamine can be liberated by the action of chitinase, but chitobiose and chitotriose account for up to 99% of the hydrolytic products (Jeuniaux, 1963). The chitinase of *Streptomyces antibioticus* has been purified by Jeuniaux (1957, 1959a) and its properties have been studied by him, as have the kinetics of chitobiases and the methods of quantitative determination of the enzymes of the chitinolytic systems. Jeuniaux has also developed a specific enzymatic method for the detection and determination of chitin. The principle of this method is the use of purified chitinase, in which other enzymatic activities have been removed. This method permits the quantitative estimation of total chitin in every kind of skeletal structure. More-over, a modification of the procedure, namely the application of pure chitinase either before or after treatment by hot alkalis, allows chitin in the free state ("free" chitin) to be distinguished from chitin bound to other macromolecules, especially to proteins in glyco-protein complexes ("masked" chitin).

Consideration of its phenomenological aspects leads zoologists to give the name of chitin not to the molecular entity defined above, but to "all cuticular or skeletal formations, more or less anhistous, and more or less resistant to chemical reagents, to putrefaction, to the knife of anatomists or to the razor blade of the microtome". Many entomologists have added to the confusion by using the term "chitinized" with the meaning of "sclerified" (Jeuniaux, 1963).

In the following pages, we shall consider the distribution of chitin and of chitinolytic enzymes along the branches of the phylogenic tree. The reader will find the details in the book of my colleague Ch. Jeuniaux (1963) from whom the following data are borrowed.

Fig. 26 shows the distribution of chitin in the phylogenic scheme proposed by Marcus. It appears that chitin biosynthesis is a primi-tive characteristic of the animal cell, present at the level of the monocellular root of Metozoa. Chitin indeed is the structural poly-saccharide of a number of metaplasmatic membranes formed by Protozoa, such as the "shells" of Thecamoebia (= Testacida) and the cyst walls of some rhizopods and ciliates. The cells of the organisms which constituted the ancestral root of porifera and

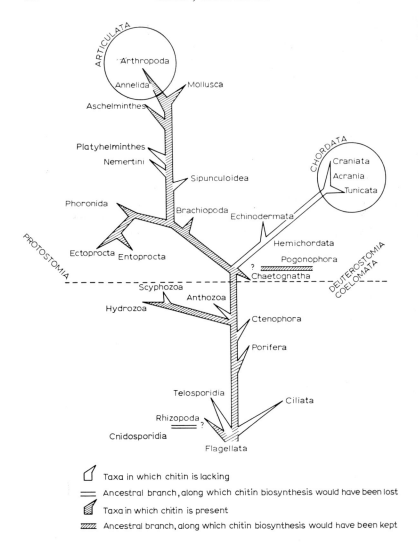

Fig. 26. Distribution of chitin and phylogenic relations proposed by Marcus (1958). (*cf*. Kerkut, 1960)

Elsevier's Scientific Publications

For information about new books in the following fields, please check square(s) and complete reverse of this card.

- ☐ PHYSICAL AND THEORETICAL CHEMISTRY
- ☐ ORGANIC CHEMISTRY
- ☐ INORGANIC CHEMISTRY
- ☐ ANALYTICAL CHEMISTRY
- ☐ BIOLOGY
- ☐ SUGAR PUBLICATIONS
- ☐ BIOCHEMISTRY
- ☐ BIOPHYSICS
- ☐ CLINICAL CHEMISTRY
- ☐ PHARMACOLOGY
- ☐ TOXICOLOGY
- ☐ PSYCHIATRY
- ☐ NEUROLOGY
- ☐ ATHEROSCLEROSIS

(please print or type)

Name: ....

Address: ....

....

....

....

POSTCARD

ELSEVIER PUBLISHING COMPANY

P.O. BOX 211

AMSTERDAM-W.
THE NETHERLANDS

Elsevier's Scientific Publications

You received this card in one of our publications. It would greatly assist us in serving you further if, when returning it for more information, you would indicate below how you heard of the book or books now in your possession. We thank you for your co-operation.

- ☐ Bookseller's recommendation
- ☐ Books sent on approval by bookseller
- ☐ Displays in bookshops
- ☐ Reviews
- ☐ Advertisements
- ☐ Personal recommendation
- ☐ References in books and journals
- ☐ Publisher's catalogue
- ☐ Circular received from publisher
- ☐ Circular received from bookseller
- ☐ Listing in a subject catalogue of bookseller

diblastic acoelomates were probably equipped with chitin-bio-synthetic systems, since chitin is actually used by Spongillidae for the construction of the gemmule walls, and by many coelenterates, especially by thecate and athecate Hydrozoa, for the building of peridermic structures (hydrorhiza, hydrocauli, hydrotheca). Chitin biosynthesis is encountered in nearly all Protostomia, the property being absent in only three taxa: Platyhelminthes, Nemertini and Sipunculoidea. As far as Deuterostomia are concerned, chitin has been found in only two very primitive and aberrant taxa, Chaetoptera and Pogonophora, the position of which, among Deuterostomia, is far from conspicuous.

It therefore seems that chitin biosynthesis, resulting from the presence of the enzyme chitin synthetase, is a primitive characteristic of the animal cell—a property which has been kept during the evolution of the ancestral forms and of most of the advanced ones in the protostomian lineage, but lost at the starting point of the deuterostomian branch.

As we have already proposed, isologous biochemical molecules other than proteins and nucleic acids will be called homologous (*indirect homology*) if the enzymes controlling their biosynthetic pathways are themselves homologous (and in the case of these enzymes, we speak of *direct homology*). Different chitins show a marked degree of homology, in that they are hydrolyzed into chitobiose by the same purified chitinase. This enzyme being highly specific of the β-glucoside link, its action confirms the presence of this link between the acetylglucosamine units. Lotmar and Picken (1950) and Rudall (1955) have demonstrated the existence of crystallographic differences between the different chitins, but as chemical analyses of α- and β-chitins have shown, the chemical constitution of these molecules does not differ; the crystallographic differences are due to distribution along more or less dense lattices.

The biosynthesis of chitin has been elucidated by experiments on *Neurospora crassa*. Uridine-diphosphate–acetylglucosamine (UDPAG) plays the role of acetylglucosamine donor. The transfer is catalyzed by a chitin synthetase (Glaser and Brown, 1957). UDPAG has been detected in the epidermis of *Maia squinado* and

Cancer pagurus (Lunt and Kent, 1961), in the hemolymphs of
Lepidoptera (Carey and Wyatt, 1960), in the wings of *Schistocerca
gregaria* (Candy and Kilby, 1962), as well as in several tissues of
vertebrates not secreting chitin (Lunt and Kent, 1961). Hence it is a
protein—the enzyme chitin synthetase, which is responsible for the
biosynthesis of chitin. This enzyme has indeed been identified in
cell-free extracts of the caterpillar of *Prodenia eridania* (Jaworski
et al., 1963) and recently in larvae of *Artemia salina* and in the
epidermis of molting blue crabs, *Calinectus sapidus* (Carey, 1965).
As we have said, consideration of the distribution of chitin along
the phylogenic tree leads to the notion of a loss of chitin synthesis
on the bifurcation of triblastic organisms. Contrary to what obtains
in Deuterostomia, chitin synthesis is preserved in Protostomia.
It is lost secondarily in a few groups, such as Platyhelminthes and
Nemertini. On the other hand, in those groups of Protostomia
which have retained chitin synthesis, the utilization of chitin is
intensified—an aspect which finds its fullest expression in Arthropoda. Where chitin synthesis is preserved along the phylogenic tree,
the product of this pathway is not modified in its chemical structure
and properties.

Chitin phylogeny is thus characterized by the loss of chitin synthesis, or, more precisely, by enzymapheresis (loss of enzymes), but,
along the branches in which its chemical nature is preserved, chitin
also shows a number of physiological and morphological radiations*. Chitin is primarily used in the formation of protective envelopes, the rigidity of which is insured by tanning of a protein combined with chitin, by calcification resulting from the deposition of
calcium carbonate, or by silicification. Such protective envelopes
of organisms are represented by the theca and cyst walls of rhizopods and ciliates, by the hydrotheca of the calyptoblastic Hydrozoa,
by the tubes of Phoronidia and Pogonophora, by the shells of in-

---

* Molecular or macromolecular units may, in phylogeny, undergo changes of
structure and/or of properties. Whether or not they undergo such changes,
they may, at a level higher than the molecular one, become involved in different
functions: this is what is called here a *physiological radiation*, examples of
which will be found on the pages 83, 84 and 162–164.

articulate Brachiopoda and of mollusks, and by the cuticles of Priapulida, Onychophora and Arthropoda. In Nematoda, chitin is used for the formation of one of the envelopes surrounding the eggs. It surrounds the latent form of life (gemmules) in *Spongilla*. Besides the function of protection of the individual, chitin in Hydrozoa and Bryozoa insures the skeletal unity of the colony. It contributes to the formation of locomotor organs, such as the loco-motor appendages of Arthropoda and the rigid setae of Annelida. Various kinds of buoyancy organ are constructed with the use of chitin, as for instance, the pneumatophores of Siphonophora, the shells of *Nautilus* and *Spirula* and the cuttlebone of *Sepia*. Chitin may also play a part in the adhesion of the organism to its substrate, as, for instance, in the cuticle of the peduncle and its ramifications in articulate Brachiopoda. At the level of the function of digestion, chitin also shows different forms of utilization. It contributes to the formation of several organs active in the capture or mastication of food (jaws and radula of gastropod and cephalopod mollusks). It constitutes the gastric shield of pelecypods, which is in contact with the "crystalline style", producing extracellularly acting en-zymes. Chitin is also used in the protection of the mucosa of the mesentera and in the formation of feces (peritrophic membrane of Arthropoda).

Another aspect of the evolution of chitinous structures is the variation in the relative proportions of free to "masked" chitin (the latter being generally implicated in glycoprotein complexes). In the most primitive groups (Protozoa, Porifera, Coelenterata), the chitinous structures are essentially built up of masked chitin. The same is true of the chitinous structures of most of the in-vertebrates, including Arthropoda, in which the proportion of free chitin is relatively low, comprizing between 5 and 17.5% of the total chitin. Two phyla seem to be exceptions to this rule: poly-chetes, the setae of which are characterized by a particularly low proportion of free chitin (only 0.5–2% of the total chitin); and, in sharp contrast, mollusks, the shells of which contain chitinous layers (*i.e.* periostracal and nacreous layers) characterized by a high proportion of free chitin (25–68% of the total chitin). Whether these different proportions of free to bound chitin influence the composition or the mechanical properties of the structures is far from clear, but they constitute biochemical taxonomic characteristics, opposing polychetes, on the one hand, and mollusks, on the other, to the other invertebrate phyla.

Chitinolysis, as stated above, is the result of the action of two enzymes: chitinase and chitobiase. The ability of animal tissues to secrete chitinase was long a controversial subject, but doubts were finally removed when the secretion of chitinase by isolated mucosa of reptiles was demonstrated by Dandrifosse in 1962. Moreover, it has been established that chitinase and chitobiase are synthetized by many unicellular organisms, such as bacteria and protozoa, and that chitinases are among the first hydrolases to be secreted outside the cells by multicellular organisms acquiring an extracellular mode of digestion (for instance, sea anemones). Chitinolytic enzymes thus appear to be part of the primitive hydrolytic equipment of heterotrophic cells.

If we trace through the branches of the phylogenic tree (Jeuniaux, 1963) the changes in the chitinolytic system (chitinase + chitobiase), we again encounter instances of enzymapheresis (loss of enzyme) involving one or other agent of the chitinolytic system, or both. The gastroderm of Cnidaria synthetizes both enzymes in large amounts, while the epidermis, rich in chitobiase, also displays a slight chitinolytic activity. The changes along the phylogenic series are different at the level of each embryonic layer. At the level of the tissues of endodermic origin, chitinase is retained in Deuterostomia as well as in Protostomia. In the chitin-eaters, the biosynthesis of chitinase is increased, while it is diminished or abolished in the non-chitinophagous forms. Among Arthropoda or Vertebrata, chitinase is only found in those species whose diet is, at least partially, formed by insects, zooplankton or fungi. With regard to chitobiase, it is preserved in the endodermic tissues of all Protostomia, while in Deuterostomia a progressive loss is observed. In the tissues of ectodermic origin, the two enzymes of the chitinolytic system are retained in Protostomia, such as Nematoda and Arthropoda, while in Mollusca only chitobiase is preserved. In Deuterostomia both enzymes are lost at the level of ectodermic tissues. These aspects of preservation or loss of chitinolytic enzymes appear to be related to physiological radiations which have occurred in the processes of digestion or of growth, among others.

In chitinophagous invertebrates, the activity of the chitinolytic system insures the liberation of acetylglucosamine, while in chitinophagous vertebrates, who do not possess chitobiase, the action of the chitinase secreted in the digestive tract no doubt permits the effective digestion of the food and the eventual absorption of chitobiose, though this has yet to be experimentally verified.

The preservation of the whole chitinolytic system at the level of the ectodermal tissues in Nematoda and Arthropoda facilitates the hatching of larvae in Nematoda, while it is taken advantage of in the process of molting in Arthropoda.

In *Beauveria bassiana*, a fungoid parasite of insects, the chitinolytic system permits the penetration of the parasite into the host.

Turning briefly to the Amphibia, we may suppose that the synthesis of chitinase and of chitobiase are related to metamorphosis

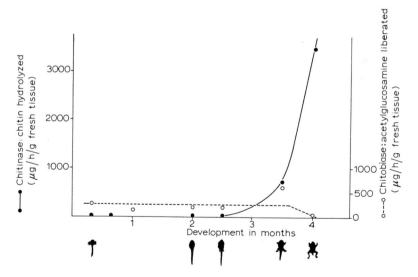

Fig. 27. Variations of chitinase and chitobiase biosynthesis, at the level of the digestive tube, throughout *Rana temporaria* L. metamorphosis. (Jeuniaux, 1963)

and represent one of its biochemical aspects. For instance, the tad-poles of *Rana temporaria*, whose food is made up of plant material, are able to digest chitin (Fig. 27). Chitinase appears with the differentiation of the stomach. During the period corresponding to the resorption of the tail and preparation for a terrestrial mode of life, a simultaneous secretion of chitinase and chitobiase is observed. Later on, the gastric mucosa of the young adult produces large amounts of chitinase, but secretes no chitobiase, or only small amounts.

One of the most remarkable physiological utilizations of the chitinolytic system is that which takes place during molting in Arthropoda. Periodically, in these animals, the old cuticle is rejected in the form of an exuvia (exuviation, or ecdysis). The pre-ecdysial period is characterized by the coming into play of molting hormones, and also by the degradation of the old cuticle by chitinolytic and proteolytic enzymes of epidermic origin. According to the terminol-ogy proposed by Drach (1939), the stages immediately following the process of exuviation in decapod Crustacea are A and B, during

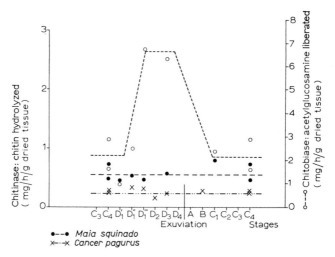

Fig. 28. Variations of chitinase and chitobiase concentrations in the epidermis of crabs, during molting cycle. Chitinases: ●−−−●, *Maia squinado*; ×—·····—×, *Cancer pagurus*. Chitobiases: ○···○, *Maia squinado*. (Jeuniaux, 1963)

which the cephalothoracic shield remains more or less soft. Calcification progresses during stages C_1 and C_2. At stage C_3 the exoskeleton is hard, but the innermost layer (membranous layer) is not yet secreted. C_4 is the period of stability corresponding to the intermolt. Stages D_1, D_2, D_3 and D_4 are the pre-exuvial stages.

Fig. 28 shows, in the case of *Maia squinado* and of *Cancer pagurus*, the changes in the biosynthesis of chitinase and of chitobiase taking place during the molting cycle. The epidermis permanently synthetizes chitinases. The exoskeleton becomes gelified at the level of the membranous layer, this gelified material being, as shown by Jeuniaux (1959b), a glycoprotein complex of chitin and proteins, and resistant to the action of the chitinases as well as of proteolytic enzymes. During the transition from the membranous layer to the gelified layer, a liberation of acetylglucosamine and of amino acids is observed, as a result of the hydrolysis of chitin and of proteins. Jeuniaux considers that glycoprotein complexes endowed with particular physical properties are embedded in a matrix of free chitin and free proteins. During stage D_2 a marked degradation of the exoskeleton is taking place. It is the membranous layer which is attacked, as stated above. The enzymatic hydrolysis of the chitin and proteins of the exoskeleton is particularly important during stages D_3 and D_4 immediately preceding exuviation. The products of this hydrolysis are resorbed by the epidermis.

In the silkworm, *Bombyx mori*, it is possible to distinguish a number of physiological ages. At the time of molting the epidermis retracts and secretes an exuvial fluid rich in chitinolytic and proteolytic enzymes. Fig. 29 shows the cyclic character of the secretion of chitinase by the epidermis. The chitinase activity is feeble or absent during the intermolt period. It appears at the beginning of each molting period and disappears immediately before ecdysis.

The construction of an exoskeleton of chitin and its degradation by the periodic appearance of chitinolytic–enzyme systems are therefore special aspects of more ancient biochemical systems and are examples of physiological radiations of these systems.

In conclusion, we may suppose that, along the branches of the phylogenic tree of invertebrates which have preserved the bio-

Bibliography p. 61

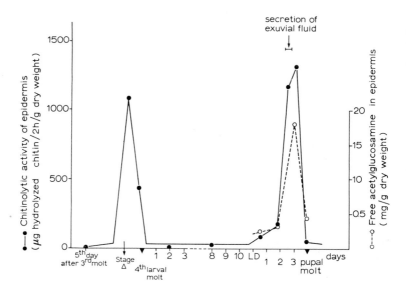

Fig. 29. Cyclic biosynthesis of chitinases by *Bombyx mori* epidermis. LD = last
defecation. (Jeuniaux, 1963)

synthesis of chitin, several attempts at protection by the use of
chitinous structures and by the utilization of chitinases at the level
of the epidermis are identifiable, the maximum advantage of these
systems having been taken by Arthropoda. Not only have the latter
preserved systems of chitin biosynthesis and chitinolysis at the level
of the epidermis, but the systems themselves are geared to cyclic
functioning under the control of hormones and forming the molecular substrate to their mode of development.

This chapter on chitin and chitinolysis, surveyed along the
branches of the phylogenic tree, shows what a variety of results
Nature has obtained without modification of structure in a biochemical component, simply by the conservation or the discon-
tinuance of protein biosynthesis, or by the introduction of the same
biosynthetic systems into a number of different physiological
complexes.

BIBLIOGRAPHY

CANDY, D. J., AND B. A. KILBY, Studies on chitin synthesis in the desert locust, *J. Exptl. Biol.*, 39 (1962) 129–140.

CAREY, F. G., Chitin synthesis *in vitro* by crustacean enzymes, *Comp. Biochem. Physiol.*, 16 (1965) 155–158.

CAREY, F. G., AND G. R. WYATT, Uridine diphosphate derivatives in the tissues and hemolymph of insects, *Biochim. Biophys. Acta*, 41 (1960) 178–179.

DANDRIFOSSE, G., La sécrétion de chitinase par la muqueuse gastrique isolée, *Ann. Soc. Roy. Zool. Belg.*, 92 (1962) 199–201.

DRACH, P., Mue et cycle d'intermue chez les crustacés décapodes, *Ann. Inst. Océanograph.*, 19 (1939) 103–106.

GLASER, L., AND D. H. BROWN, The synthesis of chitin in cell-free extracts of *Neurospora crassa*, *J. Biol. Chem.*, 228 (1957) 729–742.

HORTON, D., AND M. L. WOLFROM, Polysaccharides (excluding glycuronans, bacterial polysaccharides and mucopolysaccharides), in M. FLORKIN AND E. H. STOTZ (Eds.), *Comprehensive Biochemistry*, Vol. 5, Elsevier, Amsterdam, 1962, pp. 189–232.

JAWORSKI, E., L. WANG AND G. MARCO, Synthesis of chitin in cell-free extracts of *Prodenia eridania*, *Nature*, 198 (1963) 790.

JEUNIAUX, CH., Purification of a streptomyces chitinase, *Biochem. J.*, 66 (1957) 29 P.

JEUNIAUX, CH., Recherches sur les chitinases. II. Purification de la chitinase d'un streptomycète, et séparation électrophorétique de principes chitinolytiques distincts, *Arch. Intern. Physiol. Biochim.*, 67 (1959a) 597–617.

JEUNIAUX, CH., Sur la gélification de la couche membraneuse de la carapace chez les crabes en mue, *Arch. Intern. Physiol. Biochim.*, 67 (1959b) 516–517.

JEUNIAUX, CH., *Chitine et Chitinolyse*, Masson, Paris, 1963.

KERKUT, G. A., *Implications of Evolution* (Intern. Series of Monographs on Applied Biology, Division Zoology, Vol. 4), Pergamon, Oxford, 1960.

LOTMAR, W., AND L. E. R. PICKEN, A new crystallographic modification of chitin and its distribution, *Experientia*, 6 (1950) 57–58.

LUNT, M. R., AND P. W. KENT, Evidence for the occurrence of uridine diphosphate *N*-acetylglucosamine in crustacean tissues, *Biochem. J.*, 78 (1961) 128–134.

MARCUS, E., On the evolution of the animal phyla, *Quart. Rev. Biol.*, 33 (1958) 24–58.

RUDALL, K. M., The distribution of collagen and chitin, *Symp. Soc. Exptl. Biol.*, 9 (1955) 49–71.

ZECHMEISTER, L., AND G. TOTH, Zur Kenntnis der Hydrolyse von Chitin mit Salzsäure. I., *Ber. Deut. Chem. Gesell.*, 64 (1931) 2028–2032.

Chapter 6

Terminal Products of Nitrogen Metabolism.
Ecological and Phylogenic Aspects

It is a general belief that the nature of the nitrogen end-product of protein degradation depends, in animals, upon the abundance of water in the animal's embryonic environment. The nature of the end-product is thence linked to the ecological aspects of embryonic life. This concept is related also to the notion of the toxicity of ammonia, and it is generally believed that, while aquatic animals have a large reservoir in which to excrete the ammonia resulting from the deaminations, animals which have been able to develop a terrestrial ecology can only do this if they have acquired the means of disposing of ammonia by the use of an excretion synthesis.

Those who have undertaken comparative studies of the body fluids from the point of view of their ammonia content have been perplexed by the formation of ammonia in shed blood. As soon as the blood of mammals leaves the vessels, a rapid production of ammonia (ammoniogenesis α) occurs, lasting for about five minutes (Conway and Cooke, 1939). This is followed by a further small but progressive rise. When the curve of ammoniogenesis α is followed and extrapolated to zero time, one can calculate the levels of blood ammonia *in vivo*. Conway and Cooke (1939) have reported that in mammals and birds the curve reveals that the initial blood-ammonia level is extremely low, having a maximum value of 0.01 mg/100 ml. A similar result has been found in the case of poikilotherm vertebrates, such as frog, *Emys orbicularis*, *Clemmys leprosa*, tench and trout (Florkin, 1943). In vertebrates, therefore, the levels of blood ammonia appear to be almost negligible. This is also the case in insects, but not in the snail, lobster or crayfish. In *Helix pomatia*, for instance, the blood contains from 0.7 to 2.0 mg of ammonia per 100 ml, and in the lobster *Homarus vulgaris*, from 1.6 to 1.8 mg (Florkin and Renwart, 1939), while the value is 1.9 in the crayfish

(Florkin and Frappez, 1940). The notion of circulating ammonia has also been confirmed with respect to *Octopus* (Delaunay, 1931; Potts, 1965).

These results show that it is not the toxicity of ammonia alone which governs the general nature of excretory nitrogen products.

In aquatic animals, elimination of ammonia can be accomplished in the external liquid medium.

Sharks, lungfish, amphibia, some members of the Chelonia and the mammals dispose of ammonia by way of the biosynthesis of urea. The biochemical system used for ureogenesis in the animals mentioned above is an extension, by the addition of arginase, of the system of *de novo* synthesis of arginine.

In the system of ureogenesis, the ammonia fixation in glutamine and asparagine is completed by another ammonia-fixation reaction which occurs in the liver of the ureotelic vertebrates. Our present conception of the biochemical system operating in urea synthesis is

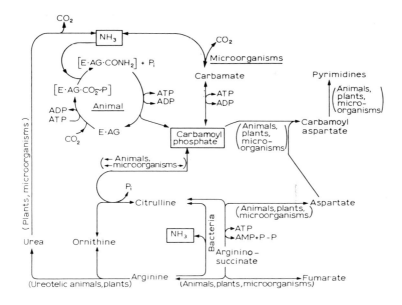

Fig. 30. Metabolic interrelationship of carbamoyl phosphate, urea cycle and pyrimidine synthesis. (Cohen and Brown, 1960)

shown in Fig. 30, in which a close relationship to pyrimidine synthesis is also apparent.

The steps of ureogenesis may be formulated as follows:

CO_2 + NH_3 + ATP (+ glutamyl derivative) → carbamoyl phosphate (1)

Carbamoyl phosphate + ornithine → citrulline (2)

Citrulline + aspartate + ATP → argininosuccinate (3)

Argininosuccinate → arginine + fumarate (4)

Arginine → ornithine + urea (5)

The enzymes catalyzing the successive steps are the following: carbamoylphosphate synthetase (1); ornithine carbamoyltransferase (2); argininosuccinate synthetase (3); argininosuccinate lyase (4); arginase (5).

Enzymes 1–4 are generally employed in the synthesis of arginine for protein synthesis as well as for phosphagen synthesis. Carbamoyl phosphate takes part in pyrimidine biosynthesis, at least in bacteria. Not all animal tissues display carbamoylphosphate synthetase activity, but other mechanisms are possible which would lead to carbamoyl aspartate, the precursor of orotic acid, and hence of pyrimidines, in animal tissues as well as in bacteria.

A proviso must be added to what has been stated above about the evolution of the system of ureogenesis as it is recognized in the liver of ureotelic vertebrates. While it is true that the enzyme carbamoylphosphate synthetase exists in bacteria, as a part of the system of arginine biosynthesis, this enzyme is not identical with the vertebrate enzyme. They both yield carbamoyl phosphate as the main end-product, but the catalytic role of the two enzymes is different. This may be regarded as a case of modification of the structure and properties of the enzyme, but we must not discard the possibility of an aspect of analogy resulting from the development of an entirely new catalytic molecule. In animal preparations, 2 moles of ATP are required for the synthesis of 1 mole of carbamoyl phosphate, while 1 mole of ATP only is required in bacterial preparations. The animal enzyme which synthetizes carbamoyl phosphate has a higher affinity for ammonia than the bacterial enzyme has, and enables it to pick up ammonia at lower concentrations than in the case of the bacterial enzyme (Cohen and Brown, 1960).

A phylogenic survey of the urea-cycle enzymes in vertebrates can be found in Cohen and Brown (1960). The elasmobranchii seem to possess all of the urea-cycle enzymes. As we have stated in regard to the bony fishes, members of the sub-class of ray-finned fish, appear to lack both carbamoylphosphate synthetase and also ornithine carbamoyltransferase activities in the liver. Another subclass of fishes, Dipnoi (lungfish), is endowed with the system of ureogenesis in the liver. Agnatha, such as hagfish and lampreys have not been studied with respect to the enzymes of the ureogenesis system. The urea-cycle enzymes have been shown to occur in the liver of all amphibians and it is of particular interest that the levels of carbamoylphosphate synthetase reflect the extent of ontogenic development. In the amphibian series, the most specialized forms show a loss of larval characteristics. A similar relationship obtains for all the enzymes during the ontogeny of the frog. The shift from ammonia to urea depends on some type of induction of the urea cycle at metamorphosis during which an enhanced level of all enzymes is observed.

Chelonian reptiles are extremely versatile and have the ability to excrete their nitrogen as ammonia, urea and/or uric acid depending upon species and environment. They seem to have all the components of the urea-cycle, while in the liver of the lizards and snakes, one cannot find the full complement of urea-cycle enzymes. This also applies to birds. It is sometimes claimed, from the graph of the nitrogen excretion of the chick, that during embryonic development the bird passes through a biochemical recapitulation excreting ammonia first, then urea, and finally uric acid. Needham et al. (1935) have shown that the urea excreted by birds derives from the effect of arginase on the arginine derived from the proteins of the yolk. Fig. 31 shows the presence or absence of the full complement of urea-cycle enzymes in certain vertebrate groups, as presented by Cohen and Brown (1960).

It certainly is difficult to connect the presence of the ureogenesis system with any aspect of water availability during embryonic life, and it would seem safe to abandon the idea. The possession

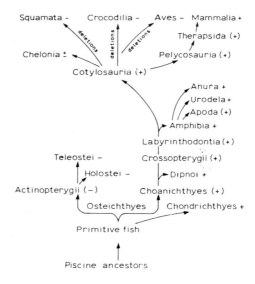

Fig. 31. Evolutionary tree of vertebrates (after Romer, 1962, modified). The assignment of + or — (after Cohen and Brown, 1960, modified), indicates the presence or absence of the full complement of urea-cycle enzymes. Signs in parentheses indicate postulated presence or absence.

of urea synthesis is better accounted for as the means of solving osmotic problems: in selachians, for instance, in the transition from fresh to sea water, the problem of equilibration with the outside medium by conserving a high concentration of small nitrogenous molecules in the body fluids, as well as in the tissues.

Smith (1936) has pointed out that it is much more correct to consider the cleidoic egg as an adaptation secondary to a retention of urea. Having developed synthesis and retention in the adult as a valuable osmotic asset, these animals "have found it convenient to enclose the embryo within a closed egg until such time as the membranes of the developing embryo can function to restrain the diffusion of the maternal gift of urea, as well as the complement added by its own metabolism. This view explains why, as an alternative to the cleidoic egg, the elasmobranchii have resorted to oviparous reproduction. Intra-uterine development will serve the purpose of protecting the young embryo against loss of urea better even than an egg-case".

Why has ureogenesis been retained in the amphibia, at least in the adult stage? We must first point out that osmoregulation is, in the terrestrial stages of amphibia, mostly an aspect of permeability as these animals do not drink. The problems of water regulation and of inorganic regulation must be considered separately not only from a pedagogic viewpoint, but also from the functional point of view, as the water and the sodium, for instance, cross the epithelium of the skin through different channels. This has been demonstrated by Schoffeniels and Tercafs (1962) by measurements of net flux of water and of electrical potential difference obtained in the isolated frog skin as affected by the antidiuretic hormone (ADH), curare, atropine and 2-PAM (pyridoxine-2-aldoxime methiodide). The discovery in Thailand by Neill (1958) of a marine frog, *Rana cancrivora*, has changed our views on the ecology of amphibia. *Rana cancrivora* has been the subject of a number of studies by Gordon *et al.* (1961), Schmidt-Nielsen and Lee (1962) and Thesleff and Schmidt-Nielsen (1962). In this species, the concentration of the blood follows the variations of the external medium, and it has been observed that, in sea water, the inorganic concentration of the blood is doubled with respect to the concentration in fresh water, the osmotic deficit being compensated by a great amount of urea, the concentration of which may reach 480 mM.

Schoffeniels and Tercafs (1966b) have performed a detailed study on the euryhaline toad, *Bufo viridis*, which adapts itself to different concentrations in the surrounding medium. Table V shows that when *Bufo viridis* is placed in solutions more concentrated than fresh water, the concentration of urea in the blood is increased, as well as that of the inorganic constituents. In *Rana temporaria* and in *Bufo bufo* only the latter constituents are increased, as shown in Table V.

What is therefore the property which enables some amphibia to live in brackish or even in sea water? If we compare the results obtained with *Rana temporaria*, *Bufo bufo* and *Bufo viridis* (including those of Gordon (1962) on *Bufo viridis*), as well as those of Gordon

TABLE V

OSMOTIC PRESSURE AND PLASMA COMPOSITION IN THREE SPECIES OF AMPHIBIANS ADAPTED TO MEDIUM OF VARIOUS CONCENTRATIONS

The calculated osmotic pressure is obtained by adding the various substances analyzed. The osmotic deficit (Δosm) represents the difference between the observed and the calculated osmotic pressure.

	Outside medium	Plasma composition							
	Osmotic pressure (mosmol/l)	Osmotic pressure (mosmol/l)		Na	Cl	K	Urea	Amino nitrogen	Δosm
		observed	calculated	(mequiv./l)			(mmol/l)		
Bufo bufo	2	213	—	145	—	4.0	10.5	—	—
	410	318	—	179	—	5.2	12.5	—	—
Rana temporaria	78	356	238.2	143.2	83.0	2.2	9.8	—	117.8
	402	510	311.4	172.0	134.0	—	4.4	—	198.6
	520	510	329.9	141.5	176.0	3.5	7.9	—	181.1
Bufo viridis	55	348	274.1	120.2	99.3	3.6	36.1	—	73.9
	144	484	358.7	145.0	118.0	—	89.0	6.7	125.3
	618	544	517.8	230.5	179.5	6.4	101.3	—	26.2
	806	778	649.0	213.0	206.0	—	218.0	12.0	129.0

et al. (1961) on *Rana cancrivora* and of Scheer and Markel (1962) on *Rana pipiens*, it appears that the species which are able to live in salt solutions containing more than 15 g NaCl/1 have a higher urea concentration (a physiological uremia) in fresh water, and that they are able to increase the blood-urea concentration when transferred to concentrated media. This adjustment in the form of increased uremia also obtains in the case of dehydration, as appears from the fact that, in dehydration, *Xenopus laevis* exhibits increased uremia up to 10 times the normal value. The same observation can be made in *Bufo bufo* and *R. temporaria*, as shown by the results given in Table VI. The most interesting finding is that in dehydrated animals the sodium content of the plasma decreases while the osmotic pressure goes up, indicating that the concentration of this ion could be of critical survival value. This could explain why the above species are unable to withstand prolonged contact with a concentrated medium which induces the sodium content of their plasma to markedly increase (Table VI). It thus appears that while the tendency of the plasma sodium to increase during dehydration can be coped with by the kidney, an increased influx of Na from the outside medium can not, and leads to a lethal increase in the sodium

TABLE VI

OSMOTIC PRESSURE AND PLASMA COMPOSITION OF TWO SPECIES OF AMPHIBIANS
PLACED IN FRESH WATER OR IN A DRY ENVIRONMENT

The dehydration amounts to about 20% of the weight.

	Osmotic pressure (mosmol/l)	Na	Cl	K	Urea (mmol/l)
			(mequiv./l)		
Bufo bufo					
fresh water	213	145	111–116	4.0	10.5
dehydrated	358–372	138.3–140		6.41–7.1	59.7–63.13
Rana temporaria					
fresh water		103.8	75.4	2.2–2.5	5.89–11.71
dehydrated		84.5–88.16	84.21–91.32	5.08–6.12	42.29–56.01

Bibliography p. 85

TABLE VII

MONTHLY VARIATIONS IN PLASMA COMPOSITION OF THE TORTOISE *Testudo hermanni* Gmelin

The concentrations in Na, K, Ca and Cl are expressed in mequiv./l, the urea concentration in mmol/l and the osmotic pressure in mosmol/l (after Gilles-Baillien and Schoffeniels, 1966)

	Na	K	Ca	Cl	Urea	Osmotic pressure measured	Osmotic pressure calculated
April 1964 (2)	167.1±3.7	4.60±0.23	4.58±0.08	134.0±12.0	103.7±9.8	467.2±19.0	415.5±24.0
May 1964 (1)	129.2	4.99	4.90	86.6	37.3	340.6	263.0
June 1964 (1)	105.2	4.25	1.50	66.8	26.6	258.5	204.4
July 1964 (3)	115.4±6.3	4.47±0.34	4.53±1.87	94.7±5.2	3.9±0.7	290.2±5.5	223.1±7.2
August 1964 (1)	135.8	4.81	5.50	108.7	12.4	322.9	267.2
September 1964 (5)	135.7±9.4	4.13±1.14	4.90±4.30	99.0±23.0	11.2±8.2	338.6±40.1	254.3±25.4
October 1964 (5)	138.3±5.2	4.21±0.60	4.81±2.11	110.3±25.5	22.7±14.1	343.7±20.5	280.3±37.1
November 1964 (5)	141.1±3.3	2.98±0.57	5.15±0.25	99.4±4.0	21.9±2.0	349.1±2.1	270.6±1.0
December 1964 (2)	155.9±1.9	3.87±0.23	6.33±1.67	124.6±13.3	31.9±2.9	404.1±61.9	313.0±21.5
January 1965 (1)	156.4	3.71	2.40	124.4	31.6	349.1	318.5
February 1965 (5)	161.6±12.4	3.03±0.49	5.32±1.08	123.5±10.2	38.4±14.5	449.3±102.3	331.7±26.1
March 1965 (5)	156.9±11.0	3.82±0.42	5.40±1.92	125.9±25.2	34.8±12.6	443.2±93.3	327.1±32.9

concentration of the plasma. In the first condition, the blood-urea concentration increases as a means of limiting the loss of water through the kidney, a function normally and otherwise performed by sodium ions. Urea also plays a part in the osmoregulation of *Testudo hermanni* J. F. Gmelin during the seasonal changes (Table VII) (Gilles-Baillien and Schoffeniels, 1966). The compensating effect of uremia in dehydration is also found in mammals. In the dromedary, dehydration is accompanied by an increased uremia (Charnot, 1960), and the same is also observed in man when primary dehydration sets in (see Florkin, 1962).

Some reptiles and the birds have lost ureogenesis. They excrete their nitrogen mostly in the form of uric acid. Uric acid "forms supersaturated solutions from which it readily separates as an almost insoluble solid. It is apparently secreted into tubular urine of birds and reptiles in a concentrated solution from which, after the reabsorption of water and perhaps of cations by the cloaca, uric acid separates as a solid. Obviously the uric acid metabolism of birds and reptiles is a great advantage for arid, terrestrial life; on the other hand it would not seem, in view of the low toxicity of urea, and the fact that other animals tolerate from 2000 to 4000 mg per cent, that the mere diffusibility of this substance in the egg would determine the type of nitrogen metabolism in the adult." (Smith, 1936).

The skin of reptiles has not been studied at the molecular level, but it is generally believed to be impermeable to water: this has been considered one of the aspects of the adaptation of reptiles to terrestrial life.

According to classical ideas, amphibia are able to resist dehydration, while reptiles avoid it by having impermeable skins. This point of view does not stand up to closer analysis. Tercafs has shown that the skin of all the reptiles he has studied is permeable to water when the external surface is bathed with Ringer's solution diluted ten times. Evaporation, on the other hand, is very reduced. This seems to present a contradiction. Tercafs concludes that this variable permeability, depending on the presence of a watery solution, or of air, at the external surface of the skin, is due to the modification of the skin structures by the presence of water. If this is true, a reptile must *lose* water in sea water, and this is what is actually observed. It remains to be determined how the marine reptiles, such as the hydrophidae, compensate for this water loss under these conditions.

What physiological changes are observed in the adaptation of the sea turtle (*Caretta caretta*) to fresh water, and of the fresh-water turtle (*Clemmys leprosa*) to sea water?

If, as done by Tercafs *et al.* (1963) and Schoffeniels and Tercafs (1966a), *Caretta* is confined in fresh water, it shows a reduction of blood osmotic pressure due to changes in inorganic constituents. The urea concentrations are variable and have no connection with the osmotic phenomena. The same applies to *Clemmys leprosa* in sea water. It is known that urea excretion by chelonians is very variable.

The system of ureogenesis is still present in the less specialized reptiles, but it is incomplete in more specialized reptiles and in birds. It is often said that these groups, having lost ureogenesis, have acquired the power of biosynthetizing uric acid with the ammonia resulting from amino acid deamination. This, however, overlooks the fact that molecules of the ammonia pool take part in the biosynthesis of purines.

In animals, uric acid is formed only by the degradation of the purine-containing nucleotides, the biosynthesis of which is accomplished through the pathway leading to inosinic acid, as shown in Fig. 32.

Even though urea is related structurally to a segment of the purine ring, it cannot be incorporated as a complete unit into nucleic acid or purine excretory product. Nevertheless, different authors have maintained that, in the snail *Helix pomatia*, uric acid, which is the main nitrogenous excretion product of this mollusk, is biosynthetized from urea in the hepatopancreas (Wolf, 1933; Baldwin, 1935; Grah, 1937). According to these authors, two molecules of urea would be combined with a molecule of tartronic acid to form a molecule of uric acid, according to the scheme previously proposed by Wiener (1902) to account for the synthesis

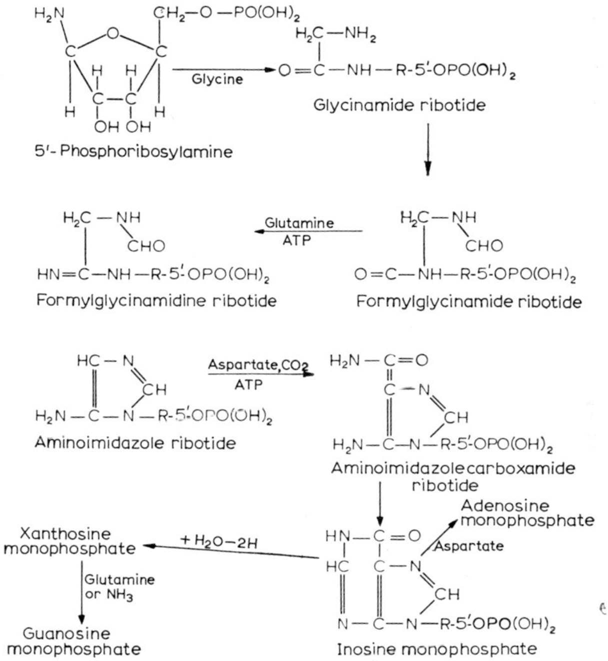

Fig. 32. Biosynthesis of the purine ring.

of uric acid in birds (Fig. 33). However, it has been shown (Bric-teux-Grégoire and Florkin, 1962) that in *Helix pomatia* the major portion of the uric acid isolated after an injection of [^{14}C]urea is localized at the level of C-6 and C-4 of uric acid. If Wiener's scheme were applicable, one would expect the activity to be located at the level of C-2 and C-8, whereas only 0.005% of the total activity injected in the form of urea is found here, the major activity being observed at C-6 and C-4; this corresponds to the positions most prominent after the administration of labelled bicarbonate (Buchanan *et al.*, 1948). The likely interpretation of these results is that the injected urea is, in the snail, decomposed in the nephridium, in the presence of the urease it contains (Baldwin and Needham, 1934; Heidermanns and Kirchner-Kühn, 1952), and that the CO_2 resulting

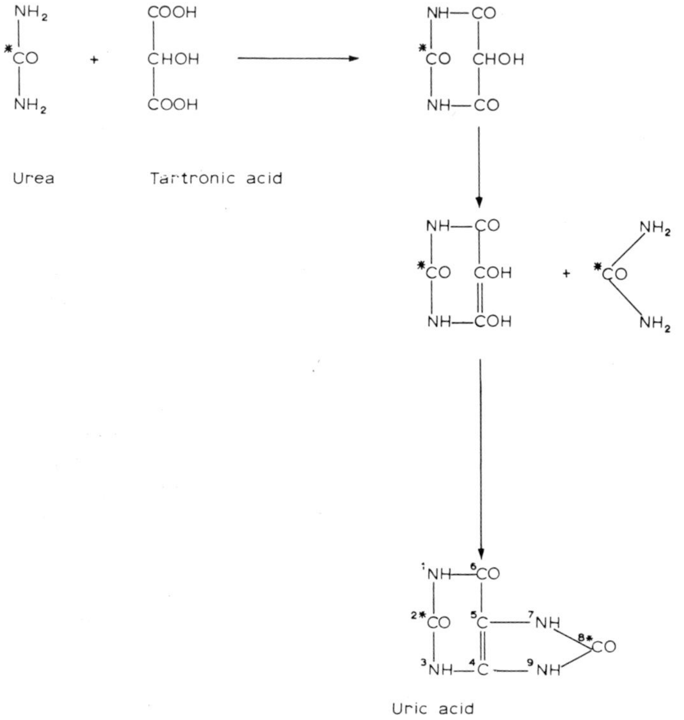

Fig. 33. Wiener's scheme for the synthesis of uric acid in birds.

from this action is used in purine-ring synthesis according to the pathway demonstrated by Buchanan *et al.* (1948). These results are in agreement with the demonstration, by Lee and Campbell (1965), of a synthesis of the purine ring, according to Buchanan's scheme, in the terrestrial snail *Otala lactea*. Heller and Jezewska (1959) have also shown that this scheme likewise accounts for the uric acid excreted by the butterfly *Antherea pernyi* and also obtains in *Escherichia coli* and in *Neurospora crassa*.

The nucleosides formed according to the pathway shown in Fig. 33 are degraded to uric acid according to the scheme of Fig. 34.

In animal tissues, the enzyme adenase is absent or nearly so, and the adenosine breakdown through the formation of adenine appears

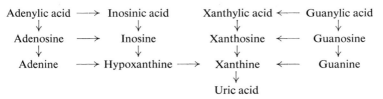

Fig. 34. From nucleosides to uric acid.

to be excluded. Small amounts of adenine can be oxidized by xanthine oxidase with the production of 8-hydroxyadenine and finally of 2,8-dihydroxyadenine (Wyngaarden and Dunn, 1957). After the deamination of guanine, xanthine is finally obtained and is, in the presence of xanthine oxidase, oxidized to uric acid. The existence of xanthine oxidase in the liver of mammals was established long ago. The enzyme is widely distributed among invertebrates. Its presence has been demonstrated in the earthworm, the pond mussel *Anodonta*, *Planorbis*, *Dytiscus*, *Gryllotalpa*, *Aeschna* as well as in the larvae of *Tenebrio* and *Limnophilus* (Florkin and Duchâteau, 1941).

The degradation of the purine ring of uric acid (purinolysis) results from the action of a system of enzymes, as shown in Fig. 35, the result of which is a conversion of the nitrogen of the purine nucleus to ammonia. This is not what obtains in *Sipunculus*, mussel, lobster and crayfish—all aquatic forms, which are provided with the whole series of the enzymes of uricolysis, and excrete ammonia as the end-product of their amino acid metabolism (Florkin and Duchâteau, 1943). Whatever its origin, ammonia is, in these creatures, simply drained into the medium. The series of enzymes of purinolysis is shortened by the loss of urease in those animals which have acquired ureogenesis, such as Selachii, Dipnoi and amphibia. In their case, urea is the end-product of both amino acid and purine metabolism. There is nevertheless an exception to this convergence in mammals. Though urea is excreted as the product of their amino acid metabolism, the chain of uricolytic enzymes is shorter in the mammals than it is in the amphibia. In general, the end-point of the purine metabolism of mammals possessing uricase

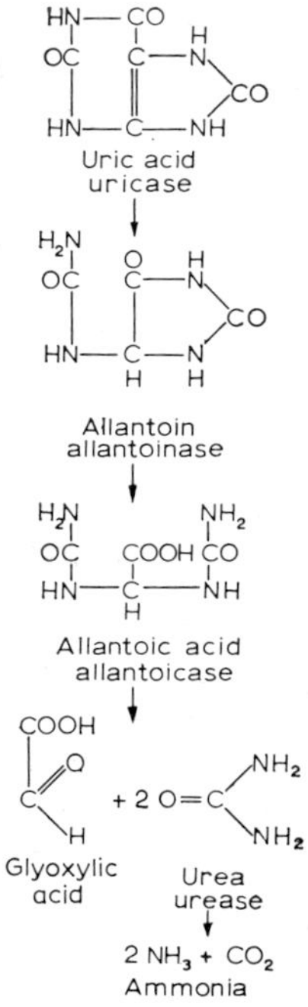

Fig. 35. The uricolytic enzyme system.

is allantoin. But in the primates, including man, the activity of uricase itself is lost or reduced, and uric acid is excreted as the end-point of their purine catabolism, together with some allantoin.

The ecological adaptation which has taken place in the uricotelic vertebrates is, therefore, a more or less marked loss of the full implement of the system of ureogenesis and a loss of the whole enzyme system of purinolysis.

In 1918, Clementi found the enzyme arginase, producing ornithine and urea and acting on arginine, in a number of invertebrates, including *Helix pomatia*, and the presence of arginase in this organism was later confirmed by Baldwin (1935). At that time, two prejudices were common among biochemists, who believed that the existence of arginase in an organism was linked with the ureotelic nature of its nitrogen metabolism, and that the snail *Helix pomatia* excreted a part of the nitrogen in the form of urea.

These prejudices led to a search for the production of urea in experiments *in vitro*, mostly accomplished with homogenates of hepatopancreas (Heidermanns, 1938). Baldwin and Needham (1934) put forward arguments according to which the urea produced in the metabolism of *Helix* derives from exogenous arginine. Bricteux-Grégoire and Florkin (1962) provided direct arguments in favor of this notion. They injected [guanido-^{14}C]arginine into the hepatopancreas of a snail, and two hours later, recovered 5% of the activity in the form of urea. The possibility can be excluded that the arginine, from which the urea is derived by the action of arginase, derives from the system of urea biosynthesis, as is the case in the ureotelic vertebrates.

Linton and Campbell (1962) studied the hepatopancreas of the land snail *Otala lactea* and found it to contain enzymes (2)–(5) of the enzyme list on p. 64, but no carbamoylphosphate synthetase.

In *Helix pomatia* hepatopancreas, Bricteux-Grégoire and Florkin (1964) were unable to detect the presence of enzymes (1) and (2). *In vitro*, however, when carbamoyl phosphate and [^{14}C]ornithine are incubated with a homogenate of hepatopancreas, labelled urea is isolated.

In conclusion, it can be said that the hepatopancreas of the land snail does not contain the complete system of ureogenesis. This also appears to apply to the snail *Lymnaea stagnalis jugularis* (Friedl and Bayne, 1966).

It has been known for a long time that the proportion of nitrogen in the form of urea, in the urine of the frog or of man, amounts to 80–85% of the total, while in the excreta of birds or snakes, uric acid accounts for approximately the same proportion. On the other hand, in the urine of *Sepia officinalis*, the ammonia nitrogen represents 65% of the total non-protein nitrogen. These are examples of clearcut cases of ureotelic and ammoniotelic nitrogen metabolism, which means that in each case, a terminal product predominate markedly. Such a conclusion is justified when it can safely be assumed that the nitrogen in the excretions analyzed represents the total nitrogen actually excreted. It would be quite misleading, for instance, to take the distribution of the nitrogenous substances present in the urine of the carp *Cyprinus carpio* as indicating the relative proportions of the different nitrogenous excretion products, since some are known to be eliminated through the gills.

Some years ago (Florkin, 1945) doubt was cast on the study of the terminal products of nitrogen metabolism through the analysis of the water in which the animals are kept, and in which they eliminate not only the products of their nephridia, but also their feces. On the other hand, some excretion products may be retained in the nephridia in an insoluble form. We must, therefore, be very cautious in the interpretation of such data as those collected in Table 112 of Albritton's *Standard Values* (1954) or in Table 22 of Prosser and Brown's *Comparative Animal Physiology* (1961).

Let us consider the case of the snail *Helix pomatia*. According to Hesse (1910), *Helix pomatia* excretes 3.85 mg of nitrogen per kg. When this observation was performed, it had long been known that the whitish content of the nephridia was for the greater part composed of uric acid, as Jacobson (1820) had established, and as many authors had confirmed. Marchal (1889), having isolated the uric acid contained in the nephridia, and after purifying this uric acid, concluded that each nephridium contained more than 7 mg of uric acid. In fact, at the end of hibernation, as Baldwin and Needham (1934) have shown, a snail's nephridium contains a mean of 32 mg of uric acid, *i.e.* about three-quarters of the dry weight of the organ. Uric acid is mostly excreted in its acid form, a feeble portion being in the form of urates (Heidermanns, 1953).

In the tables referred to above, the data concerning *Helix pomatia* are those first published by Delaunay (1927) and made generally known later by the publications of Needham (1935) and of Baldwin (1947). Delaunay analyzed a "water extract" of the nephridia, prepared in such a way that, as pointed out by Jezewska *et al.* (1963), about 80% of the purine compounds were left non-extracted. Delaunay considers that the snail has two kinds of excreta: solid ones composed mainly of purines, and liquid ones, consisting mainly of ammonia and urea.

To collect the "liquid excreta", Delaunay kept the snails partially immersed in distilled water and determined the nitrogenous compounds given out in this water. This he called "liquid excretion", but his figures were placed in the tables referred to above and were considered as representing the excreta of *Helix*. Jezewska *et al.* (1963) proceeded in a different way in their study of the nitrogen compounds in snail's excretions. The hibernating operculated snails were stored in a refrigerator, at 4° C. In the spring, they were transferred to room temperature and after awakening, were fed with lettuce and cauliflower leaves in glass beakers, the bottoms of which were covered with a layer of 2–4 mm of water. The snails remained on the walls of the beakers, where they deposited their feces as well as their nephridial excreta. These were easily distinguished. The yellowish nephridial excreta were collected during April, May and June, and air-dried.

During hibernation as well as during the active period, snails were dissected and their nephridia were isolated. The content of each nephridium was rinsed out with distilled water into a 100-ml flask. A saturated solution of lithium carbonate was added to the accompaniment of shaking and moderate heating until all concrements were dissolved. The air-dried nephridial excreta were treated in the same way. The results of the analyses performed on the dissolved excreta are shown in Table VIII. They show that 90% of the total nitrogen in the excreta and in the contents of the nephridia consisted of uric acid, xanthine and guanine. Urea, ammonia and allantoin were not found. During the feeding period, the total N content in the nephridia was only 50% of that in hibernating snails.

These data allow us to explain the nature of the erroneous data still often printed. Delaunay's "water extract" represented only 522 mg N per 100 g of nephridia of feeding snails, while the figures of Jezewska *et al.*, recalculated on the same basis, represent a value of about 10.000 mg. Consequently the evaluation of urea, ammonia and amino acids by Delaunay amounts to only 1% of the total N of the excreta. Such small amounts have probably been neglected in Table VIII, the data of which were obtained on small samples. The purinotelic character of nitrogen metabolism in the snail is obvious, and the proportions in the excreta of end-products other than uric acid, guanine and xanthine, amount to mere traces.

Bibliography p. 85

TABLE VIII

NITROGEN COMPOUNDS IN NEPHRIDIUM AND NEPHRIDIUM EXCRETA IN THE SNAIL *Helix pomatia*

(Jezewska et al., 1963)

The average values of two determinations are given. The difference between the determinations did not exceed 0.5 mg per nephridium or per 100 mg of excreta. The results are presented in mg N per nephridium or per 100 mg of excreta, and in per cent of total nitrogen

Material	Expt. No.	N total (mg)	Uric acid (mg)	Uric acid (%)	Xanthine (mg)	Xanthine (%)	Guanine (mg)	Guanine (%)	Total purines (mg)	Total purines (%)	Unknown comp. (mg)	Unknown comp. (%)	N accounted for (mg)	N accounted for (%)
Nephridium during hibernation	1	43.9	25.0	56.9	8.6	19.5	5.8	13.2	39.5	90.0	0.0	0.0	39.4	90.0
	2	38.1	24.3	63.7	4.8	12.5	5.0	13.1	34.1	89.5	0.0	0.0	34.1	89.5
	3	59.4	44.6	75.0	4.8	8.0	5.9	9.9	55.3	93.0	0.0	0.0	55.3	93.0
average		47.1	31.3	66.4	6.0	12.7	5.6	11.8	42.9	91.0	0.0	0.0	42.9	91.0
during the feeding period	4	21.9	6.0	27.3	8.1	36.9	4.2	19.1	18.3	83.5	3.3	15.0	21.6	98.6
	5	22.1	5.0	22.6	8.7	39.3	5.3	24.0	19.0	85.9	3.2	14.0	22.2	100.4
	6	26.3	11.1	42.2	8.4	31.9	3.6	13.6	23.2	88.1	1.4	5.3	24.6	93.5
	7	15.1	7.8	51.6	3.6	23.8	3.6	23.8	15.0	99.3	0.0	0.0	15.0	99.3
average		21.3	7.5	35.2	7.2	33.9	4.2	19.7	18.9	88.7	1.9	8.6	20.8	97.6
Excreta first after hibernation	8	33.6	21.7	64.6	12.6	37.5	0.0	0.0	34.3	102.1	0.0	0.0	34.3	102.1
	9	33.1	12.8	38.7	12.0	36.3	3.8	11.5	28.6	86.4	2.1	6.3	30.7	92.7
	10	34.7	15.8	45.5	8.8	25.3	7.4	21.3	32.0	92.2	2.1	6.1	34.1	98.3
	11	31.1	14.1	45.3	9.0	28.9	6.9	22.2	30.0	96.5	0.5	1.6	30.5	98.0
average		33.0	14.2	43.0	9.9	30.0	6.0	18.2	30.2	91.5	1.5	4.5	31.7	96.1

Nitrogen of spans the Uric acid, Xanthine, and Guanine columns.

If we recapitulate the aspects mentioned above with respect to the terminal products of nitrogen metabolism in vertebrates, we conclude that these animals are either ammoniotelic, ureotelic or uricotelic. When ammonia can be disposed of by a rapid elimination in the external medium, the ammoniotelic type obtains. If not, the type is either ureotelic or uricotelic. We have recognized, in the ureotelic type, adaptive aspects linked to osmoregulation, while the uricotelic type is in relation to life in very dry media. All these types of nitrogen metabolism converge, in vertebrates, in avoiding a circulation of ammonia in the body and we may tentatively propose that ammonia is eliminated in relation to its depolarizing action at the level of cell membranes, in the presence of relatively small concentrations of calcium, as is the case in vertebrates. On the contrary, owing to their high calcium content, crustaceans and molluscs can withstand the presence of more ammonia. The complete ureogenesis system insures the synthesis of arginine $de\ novo$ from carbon dioxide and ammonia and the partial transformation of arginine into urea and ornithine.

But there are definite examples of ureotelism among invertebrates. One example is observed in the earthworm $Lumbricus\ terrestris$ during starvation. Although being normally ammoniotelic, the earthworm excretes as much as 90% of its non-protein nitrogen in the form of urea, during starvation (Cohen and Lewis, 1949; Needham, 1957). The complete array of enzymes of the ureogenesis system, insuring $de\ novo$ synthesis of arginine and urea from CO_2 and NH_3 has now been demonstrated in the earthworm gut tissue and it has also been shown that they function $in\ vivo$ to synthetize $de\ novo$ both arginine and urea (Bishop and Campbell, 1965). During starvation, the earthworm $Lumbricus\ terrestris$ becomes truly ureotelic. The increase of urea excretion is due to an increase in the level of the urea-cycle enzymes. As they are normally ammoniotelic, the earthworms do not seem to undergo any deleterious influence of ammonia. The amount of urea excreted seems to depend upon the amount of metabolic water available as inanition increases the excretion of urea. The importance of the shift from the ammoniotelic to the ureotelic metabolism probably lies in osmoregulatory aspects and more data on the influence of desiccation on the type of nitrogen metabolism in earthworms should become available.

Another example is provided by the terrestrial planarian *Bipalium kewense*, in which the system of ureogenesis has been identified and its presence confirmed by isotopic incorporation data (Campbell, 1965). There is also evidence in favour of the existence of the system in the different classes of flatworms (Campbell and Lee, 1963; Agosin and Repetto, 1963).

Increased knowledged has therefore shown that the presence of the system of ureogenesis by the extension, through the presence of arginase, of the system for the *de novo* synthesis of arginine, is much more widespread than was believed previously. In the light of the important facts discovered by Campbell and his collaborators, ureogenesis cannot be considered any more in the field of animal phylogeny, as being an invention of the ureotelic vertebrates (see Florkin, 1966).

The extension leading to the *de novo* production of urea (to be distinguished from the action of arginase on exogenous arginine) is not the only form of evolution of the system of arginine biosynthesis. As said above, some of the components of the system may be lost, with definite nutritional consequences, as for instance, is shown by the lack of ornithine carbamoyltransferase in insects (Porembska and Mochnacka, 1964), the result of this being that arginine is an essential amino acid in these animals. Ornithine can partially replace arginine in their diet (Hinton, 1959) as they possess enzymes (3) and (4) of the system of arginine *de novo* biosynthesis (p. 64).

None of the enzymes (1), (2), (3) or (4) are present in the liver of the chick, but the kidney possesses enzymes (3) and (4) (Tamir and Ratner, 1963a). This accounts for the ability of the chick to utilize citrulline for growth in place of arginine (Klose and Almquist, 1940). As the uricotelic vertebrates have no possibility of synthetizing citrulline (reaction 2 of the system), ornithine accumulates as it cannot be re-utilized for urea formation. Uricotelic vertebrates use glycine for detoxication to a very limited extent only (Smith, 1958). This appears as being due to the large demand for purine synthesis and subsequent uric acid excretion. This accounts for the participation of ornithine for detoxication in birds by the synthesis of ornithuric acid (Tamir and Ratner, 1963b). The fact that ornithuric acid is also the major detoxication product in snakes and lizards (Smith, 1958) is in keeping with the explanation proposed above.

Evidence showing that ammonia can be used for the synthesis of arginine, *de novo*, from CO_2 and NH_3, in a system involving the ornithine cycle, is now available, in many plant organisms, mono-cellular or multicellular, even if arginase is lacking, as there are other mechanism available for returning the citrulline to the ornithine cycle without production of urea (Literature : see Ratner, 1954).

The system of arginine biosynthesis as found in animals appears as being derived from this ancient biosynthetic system by the acquisition of a different carbamoylphosphate synthetase. This system, by an extension consisting of the possession of the enzyme arginase, provides urea, in the absence of urease. When urease exists, the end-point is ammonia.

The ammonia resulting from the catabolism of amino acids, either by deamination or by transamination, is either directly excreted if conditions permit this, or enters different pathways including those leading to glutamine and asparagine, to arginine and to purines. The system of ureogenesis as found in Protozoa*, in the earthworms, in the flatworms, and in the ureotelic vertebrates, results from the maintenance of the whole system of arginine biosynthesis supplemented by the presence of arginase in organisms deprived of urease. Ureogenesis appears in a number of cases as being connected with osmoregulatory aspects. This can be considered as a physiological radiation of the system of arginine biosynthesis. In uricotelic animals, purine biosynthesis disposes of an increased proportion of excreted nitrogen. This again appears as being a physiological radiation of the system of purine synthesis.

---

* Arguments have been provided in favour of the existence of the ureogenesis system in *Tetrahymena* (Hogg and Elliott, 1951; Wu and Hogg, 1952). Not only should this notion lead to more complete studies, but the nature of the carbamoylphosphate synthetase eventually involved, should be defined.

Bibliography p. 85

In vertebrates, either ammoniotelic, ureotelic or uricotelic, the systems contributing to nitrogen excretion insure a lack of circulating ammonia in the blood. We have tentatively linked this aspect with the depolarizing effect of ammonia in the absence of a relatively high concentration of calcium. This elimination appears again as being a physiological radiation, either of arginine biosynthesis or of purine biosynthesis. In some higher plants, ammonia is accumulated in large amounts in the form of glutamine and of asparagine.

When considering animal phylogeny, we may conclude that the differences in the terminal products of nitrogen metabolism are not the result of any change in structure or properties of the metabolic systems, but depend on the relative levels of concentration or activities of the enzymes involved (system of arginine biosynthesis, arginase, system of purine biosynthesis, system of purinolysis).

BIBLIOGRAPHY

AGOSIN, M., AND Y. REPETTO, Studies on the metabolism of *Echinococcus granulosus*. VII. Reactions of the tricarboxylic acid cycle in *E. granulosus* scolices, *Comp. Biochem. Physiol.*, 8 (1963) 245–261.

ALBRITTON, E. C., *Standard Values in Nutrition and Metabolism*, Saunders, Philadelphia, 1954.

BALDWIN, E., Problems of nitrogen catabolism in invertebrates. III. Arginase in the invertebrates, with a new method for its determination, *Biochem. J.*, 29 (1935) 252–262.

BALDWIN, E., *Dynamic Aspects of Biochemistry*, Cambridge University Press, London, 1947, 457 pp.

BALDWIN, E., AND J. NEEDHAM, Problems of nitrogen catabolism in invertebrates. I. The snail (*Helix pomatia*), *Biochem. J.*, 28 (1934) 1372–1392.

BISHOP, S. H., AND J. W. CAMPBELL, Arginine and urea biosynthesis in the earthworm *Lumbricus terrestris*, *Comp. Biochem. Physiol.*, 15 (1965) 51–71.

BRICTEUX-GRÉGOIRE, S., AND M. FLORKIN, Les excreta azotés de l'escargot *Helix pomatia* et leur origine métabolique, *Arch. Intern. Physiol. Biochim.*, 70 (1962) 496–506.

BRICTEUX-GRÉGOIRE, S., AND M. FLORKIN, Recherche des enzymes du cycle de l'uréogenèse chez l'escargot *Helix pomatia* L., *Comp. Physiol. Biochem.*, 12 (1964) 55–60.

BUCHANAN, J. M., J. C. SONNE AND A. M. DELLUVA, Biological precursors of uric acid. II. The role of lactate, glycine, and carbon dioxide as precursors of the carbon chain and nitrogen atom 7 of uric acid, *J. Biol. Chem.*, 173 (1948) 81–98.

CAMPBELL, J. W., Arginine and urea biosynthesis in the land planarian: its significance in biochemical evolution, *Nature*, 208 (1965) 1299–1301.

CAMPBELL, J. W., AND T. W. LEE, Ornithine transcarbamylase and arginase activity in flatworms, *Comp. Biochem. Physiol.*, 8 (1963) 29–38.

CHARNOT, Y., Répercussion de la déshydratation sur la biochimie et l'endocrinologie du dromadaire. *Trav. Inst. Sci. Chériffien, Ser. Zool.*, 20 (1960) 1–168.

CLEMENTI, A., Ricerche sull'arginasi. V. Sulla presenza dell'arginasi nell'organismo di qualche invertebrati. *Atti Reale Accad. Lincei, Rend.*, 27 (1918) 299–302.

COHEN, P. P., AND G. W. BROWN JR., Ammonia metabolism and urea biosynthesis, in M. FLORKIN AND H. S. MASON (Eds.), Comparative Biochemistry, Vol. 2, Academic Press, New York, 1960, pp. 161–244.

COHEN, S., AND H. B. LEWIS, The nitrogenous metabolism of the earthworm (*Lumbricus terrestris*), *J. Biol. Chem.*, 180 (1949) 79–91.

CONWAY, E. J., AND R. COOKE, Blood ammonia, *Biochem. J.*, 33 (1939) 457–478.

DELAUNAY, H., *Recherches Biochimiques sur l'Excrétion Azotée des Invertébrés*, Thèse Sci. Nat., Paris, 1927, pp. 1–196.

DELAUNAY, H., L'excrétion azotée des invertébrés, *Biol. Rev.*, 6 (1931) 265–301.

FLORKIN, M., Sur l'ammoniaque sanguine des vertébrés poecilothermes, *Arch. Intern. Physiol.*, 53 (1943) 117–120.

FLORKIN, M., *L'Évolution du Métabolisme des Substances Azotées chez les Animaux*, Masson, Paris, 1945, pp. 1–66.

FLORKIN, M., *Introduction Biochimique à la Chirurgie*, Masson, Paris, 1962, pp. 1–248.

FLORKIN, M., *Aspects Biochimiques de l'Adaptation et de la Phylogénie*, Masson, Paris, 1966.

FLORKIN, M., AND G. DUCHÂTEAU, Sur la distribution zoologique de la xanthine-oxydase, *Bull. Classe Sci. Acad. Roy. Belg.*, 27 (1941) 174–178.

FLORKIN, M., AND G. DUCHÂTEAU, Les formes du système enzymatique de l'uricolyse et l'évolution du catabolisme purique chez les animaux, *Arch. Intern. Physiol.*, 53 (1943) 267–307.

FLORKIN, M., AND G. FRAPPEZ, Concentration de l'ammoniaque, *in vivo* et *in vitro*, dans le milieu intérieur des invertébrés. III. Ecrevisse, hydrophile, dytique, *Arch. Intern. Physiol.*, 50 (1940) 197–202.

FLORKIN, M., AND H. RENWART, Concentration de l'ammoniaque, *in vivo* et *in vitro*, dans le milieu intérieur des invertébrés. II. Escargot et homard, *Arch. Intern. Physiol.*, 49 (1939) 127–128.

FRIEDL, F. E., AND R. A. BAYNE, Ureogenesis in the snail *Lymnaea stagnalis jugularis*, *Comp. Biochem. Physiol.*, 17 (1966) in print.

GILLES-BAILLIEN, M., AND E. SCHOFFENIELS, Variations saisonnières dans la composition du sang de la tortue grecque *Testudo hermanni* J. F. Gmelin, *Ann. Soc. Roy. Zool. Belg.*, (1966) in print.

GORDON, M. S., Osmotic regulation in the green toad *Bufo viridis*, *J. Exptl. Biol.*, 39 (1962) 261–270.

GORDON, M. S., K. SCHMIDT-NIELSEN AND H. M. KELLY, Osmotic regulation in the crab-eating frog (*Rana cancrivora*), *J. Exptl. Biol.*, 38 (1961) 659–678.

GRAH, H., Untersuchungen über die Wirkungsweise des Harnsäure bildenden Fermentes bei *Helix pomatia*, *Zool. Jahrb.* (*Physiol.*), 57 (1937) 355–372.

HEIDERMANNS, C., Der Exkretstoffwechsel der wirbellosen Tiere, *Naturwissenschaften*, 26 (1938) 263.

HEIDERMANNS, C., Der Anteil der Urate an der kristallinen Harnsäureexkrementen, *Naturwissenschaften*, 40 (1953) 403–404.

HEIDERMANNS, C., AND I. KIRCHNER-KÜHN, Über die Urease von *Helix pomatia*, *Z. vergleich. Physiol.*, 34 (1952) 166–178.

HELLER, J., AND M. M. JEZEWSKA, The synthesis of uric acid in the Chinese tussur moth (*Antheraea pernyi*), *Bull. Acad. Polon. Sci.*, 7 (1959) 1–4.

HESSE, P., Zum Hungerstoffwechsel der Weinbergschnecke, *Z. allgem. Physiol.*, 10 (1910) 273–340.

HINTON, T., Miscellaneous nutritional variations, environmental and genetic, in *Drosophila*, *Ann. N. Y. Acad. Sci.*, 77 (1959) 366–372.

HOGG, J. F., AND A. M. ELLIOTT, Comparative amino acid metabolism of *Tetrahymena geleii*, *J. Biol. Chem.*, 192 (1951) 131–139.

JACOBSON, L., cited by WOLF (1933).

JEZEWSKA, M. M., B. GORZKOWSKI AND J. HELLER, Nitrogen compounds in snail, *Helix pomatia*, excretion, *Acta Biochim. Polon.*, 10 (1963) 55–65.

KLOSE, A. A., AND H. J. ALMQUIST, The ability of citrulline to replace arginine in the diet of the chick, *J. Biol. Chem.*, 135 (1940) 153–155.

LEE, T. W., AND J. W. CAMPBELL, Uric acid synthesis in the terrestrial snail *Otala lactea*, *Comp. Biochem. Physiol.*, 15 (1965) 457–468.

LINTON, S. N., AND J. W. CAMPBELL, Studies on urea cycle enzymes in the terrestrial snail, *Otala lactea, Arch. Biochem. Biophys.*, 97 (1962) 360–369.

MARCHAL, P., Contribution à l'étude de la désassimilation de l'azote. L'acide urique et la fonction rénale chez les invertébrés, *Mém. Soc. Zool. France*, 3 (1889) 31–87.

NEEDHAM, A. E., Components of nitrogenous excreta in the earthworms *Lumbricus terrestris* L. and *Eisenia foetida* (Savigny), *J. Exptl. Biol.*, 34 (1957) 425–446.

NEEDHAM, J., Problems of nitrogen catabolism in invertebrates. II. Correlation between uricotelic metabolism and habitat in the phylum Mollusca, *Biochem. J.*, 29 (1935) 238–251.

NEEDHAM, J., J. BRACHET AND R. K. BROWN, The origin and fate of urea in the developing hen's egg, *J. Exptl. Biol.*, 12 (1935) 321–336.

NEILL, W. T., The occurrence of amphibians and reptiles in saltwater areas, and a bibliography, *Bull. Marine Sci. Gulf Caribbean*, 8 (1958) 1–97.

POREMBSKA, Z., AND I. MOCHNACKA, The ornithine cycle in *Celerio euphorbiae*, *Acta Biochim. Polon.*, 11 (1964) 113–119.

POTTS, W.T.W., Ammonia excretion in *Octopus dofleini, Comp. Biochem. Physiol.*, 14 (1965) 339–355.

PROSSER, C. L., AND F. A. BROWN JR., *Comparative Animal Physiology*, Saunders, Philadelphia, 2nd ed., 1961, 688 pp.

RATNER, S., Urea synthesis and metabolism of arginine and citrulline, *Advan. Enzymol.*, 15 (1954) 319–387.

ROMER, A. S., *The Vertebrate Body*, 3rd ed., Saunders, Philadelphia, 1962, 627 pp.

SCHEER, B. T., AND R. P. MARKEL, The effect of osmotic stress and hypophysectomy on blood and urine urea levels in frogs, *Comp. Biochem. Physiol.*, 7 (1962) 289–297.

SCHMIDT-NIELSEN, K., AND P. LEE, Kidney function in the crab-eating frog (*Rana cancrivora*), *J. Exptl. Biol.*, 39 (1962) 167–177.

SCHOFFENIELS, E., AND R. R. TERCAFS, Potential difference and net flux of water in the isolated amphibian skin, *Biochem. Pharmacol.*, 11 (1962) 769–778.

SCHOFFENIELS, E., AND R. R. TERCAFS, Adaptation à l'eau douce d'un reptile marin, *Caretta caretta* L., et d'un reptile d'eau douce, *Clemmys leprosa* L., à l'eau de mer, *Ann. Soc. Roy. Zool. Belg.*, (1966a) in print.

SCHOFFENIELS, E., AND R. R. TERCAFS, L'osmorégulation chez les batraciens, *Ann. Soc. Roy. Zool. Belg.*, (1966b) in print.

SMITH, H. W., The retention and physiological role of urea in the Elasmobranchii, *Biol. Rev.*, 11 (1936) 49–82.

SMITH, J. N., Comparative detoxication. V. Conjugation of aromatic acids in reptiles: formation of ornithuric acid, hippuric acid and glucuronides, *Biochem. J.*, 69 (1958) 509–516.

TAMIR, H., AND S. RATNER, Enzymes of arginine metabolism in chicks, *Arch. Biochem. Biophys.*, 102 (1963a) 249–258.

TAMIR, H., AND S. RATNER, A study of ornithine, citrulline and arginine synthesis in growing chicks, *Arch. Biochem. Biophys.*, 102 (1963b) 259–269.

TERCAFS, R. R., E. SCHOFFENIELS AND G. GOUSSEF, Blood composition of a sea turtle, *Caretta caretta* L., reared in fresh water, *Arch. Intern. Physiol. Biochim.*, 71 (1963) 614–615.

THESLEFF, S., AND K. SCHMIDT-NIELSEN, Osmotic tolerance of the muscles of the crab-eating frog, *Rana cancrivora*, *J. Cellular Comp. Physiol.*, 59 (1962) 31–34.

WIENER, H., Über synthetische Bildung der Harnsäure im Tierkörper, *Hofmeister's Beitr.*, 2 (1902) 42–85.

WOLF, G., Die physiologische Chemie der nephridialen Stickstoffausscheidung bei *Helix pomatia* L., unter besonderer Berücksichtigung der Einflüsse des Sommer- und Winterstoffwechsels, *Z. Vergleich. Physiol.*, 19 (1933) 1–37.

WU, CH., AND J. F. HOGG, The amino acid composition and nitrogen metabolism of *Tetrahymena geleii*, *J. Biol. Chem.*, 198 (1952) 753–764.

WYNGAARDEN, J. B., AND J. T. DUNN, 8-Hydroxyadenine as the intermediate in the oxidation of adenine to 2,8-dihydroxyadenine by xanthine oxidase, *Arch. Biochem. Biophys.*, 70 (1957) 150–156.

Chapter 7

Hemolymph Osmotic Effectors in Insect Phylogeny

Hemolymph is the only extracellular fluid in insects. They have given up the physiological association between the respiratory and the circulatory systems, the distribution of oxygen to all cells being insured by the tracheal system. Insects are therefore not bound to the maintenance of a definite blood volume, and they can rely on blood water to insure their survival in dry media. In spite of variations in blood volume, they can regulate the osmotic pressure in the hemolymph by changing the amino acid concentration. The aminoacidemia is high and the nitrogenous components of hemolymph are mainly made up of the components of the amino acid pool. The proteins of insect hemolymph lack the oncosmotic and nutritive components of mammalian plasma; they are mainly enzymes. The hemolymph of insects therefore is a matrix—a nutritive medium for the cells, undergoing changes of volume accompanied by osmotic regulation accomplished through changes in aminoacidemia.

By its nature as a repository of a number of reserve materials, in constant exchange relationship with the fat body, hemolymph is well suited to the life of organisms in which feeding is interrupted during certain phases, or during diapause related to factors in the environment, or in the course of ecological adaptations corresponding to different periods of development.

A concept which has been widely promoted for some time is that the inorganic composition of the medium of the cells must conform to definite relative proportions of Na, K, Ca and Mg. In his book, *An Introduction to Comparative Biochemistry* (3rd ed., 1948) Baldwin devotes eight pages to this concept and writes: "instead of being surprised that the bloods of different animals resemble each other so closely, we must realise that it could not have been otherwise. The composition of the blood has remained the same because the conditions under which life is possible have remained the same."

Bibliography p. 113

In another book, *The Nature of Biochemistry* (1962), the same author again underlines the concept. "Even the cells and organs of animals whose ancestors, like our own, became independent of the sea many millions of years ago, cannot tolerate for long any appreciable departure from the normal, sea-water-like composition of the blood as far as Na^+, K^+ and Ca^{2+} are concerned. This necessary internal constancy is something that has to be maintained." Insect hemolymph contradicts this statement.

It is a delicate matter to discuss the phylogeny of the concentrations of the constituents of organisms, because they reflect steady states, often of an intricate nature.

The osmotic pressure of the hemolymph of insects is generally somewhat higher than that of mammalian blood*. The values obtained by different authors, and carefully compiled by Sutcliffe (1963), show that the osmotic pressure, expressed in terms of freezing point lowering, generally ranges from -0.5 to $-0.9°$ C. Minimal values have been obtained in the case of *Ephemera danica* larvae ($-0.504°$ C), of three Trichoptera larvae (-0.38 to $-0.455°$ C) and of *Tipula montium* larvae ($-0.443°$ C). Higher values have been observed in the larvae of *Popillia japonica* ($-1.03°$ C) and *Ephestia kühniella* ($-1.130°$ C). The high values observed during pupal life in some Lepidoptera are not surprising, owing to the increasing amount of hydrolytic products resulting from histolysis.

In contrast to the blood of vertebrates, the sum of the inorganic cations and anions does not account for the total osmotic pressure. Free amino acids, organic acids and other organic molecules play an important role as osmolar effectors, especially in the most specialized endopterygote orders.

It is well known that, in most other animal phyla, the osmotic pressure of the body fluid is insured by inorganic constituents, among which sodium is generally the main cation, and chloride the main anion. The situation is more complicated and sometimes entirely different in the case of insects.

As Sutcliffe (1962, 1963) has pointed out, and as it appears from

---

* Pages 89–114 were written with the collaboration of my colleague Dr. Ch. Jeuniaux.

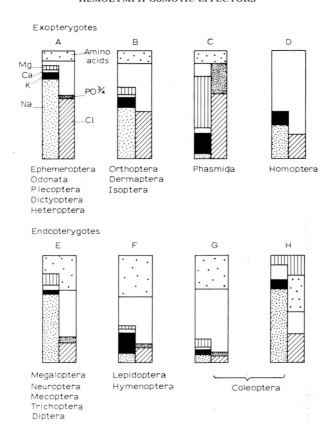

Fig. 36. Osmotic effects of components illustrated as percentages of the total osmolar concentration of hemolymph in pterygote insects. Each block in the figure is visualized as two vertical sections, each section representing 50% of the total osmolar concentration. The percentage contributions of cations are illustrated in the left-hand section, with sodium at the base (stippled), followed by potassium (black area), calcium (white area) and magnesium (vertical stripes). Anions are illustrated in the right-hand section, with chloride at the base (oblique stripes) followed by inorganic phosphate (fine stippling). Where possible, free amino acids are illustrated in equal proportions in both sections (coarse stippling). The large blank area in each block represents the proportion of the total osmolar concentration that must be accounted for by other components of the hemo-lymph. (Sutcliffe, 1963)

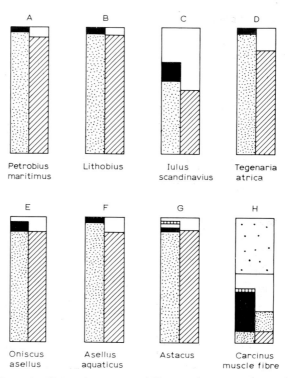

Fig. 37. Osmotic effects of components illustrated as percentages of the total osmolar concentration of the blood in: *A*, an apterygote insect; *B*, a chilopod; *C*, a diplopod; *D*, an arachnid; *E–G*, crustaceans. *H* illustrates the osmotic effects of components in the muscle fiber of *Carcinus maenas*. Conventions as in Fig. 36. (Sutcliffe, 1963)

examination of Figs. 36 and 37, the participation of inorganic cations and anions in the osmotic pressure of the hemolymph tends to decrease with the evolutionary level of the insect. Among the most primitive insects (Apterygota), *Petrobius maritimus* shows a hemolymph composition very similar to that of other arthropods, with the nearly exclusive participation of Na and Cl as osmotic effectors (Lockwood and Croghan, 1959). In the most primitive pterygote insects, all of which are exopterygotes (Ephemeroptera, Odonata, Dictyoptera, Heteroptera, and to a lesser extent, in Orthoptera, Isoptera and Dermaptera), the sum of the four cations accounts for nearly half of the osmotic pressure, Na playing the principal role, while the concentrations of K, Ca and Mg are very low. In these orders, chloride is the main anion, inorganic phosphates and organic molecules being in low concentration. In these insects, the situation is not very different from that found in other animals, and Sutcliffe (1963) has suggested that "haemolymph with this type of composition represents the basic type of haemolymph in pterygote insects."

In the Cheleutoptera (= Phasmoptera), the situation is very similar, but Mg takes the place of Na as the principal osmotic effector, and inorganic phosphates are more abundant.

A third case is represented by the following endopterygote orders: Megaloptera, Neuroptera, Mecoptera, Trichoptera and Diptera. The sum of the cations is again responsible for nearly half of the osmotic pressure, with Na as the principal effector but chloride is of minor importance and is partially replaced by amino acids and other, small organic molecules.

In Lepidoptera, Hymenoptera and many Coleoptera, the importance of cations, as well as that of chloride, is considerably reduced, organic molecules playing the main role as osmolar effectors. These groups, in which the highest values of amino acid participation are found, are also recognized by Duchâteau et al. (1953) as being highly specialized through the existence of very low values of the Na index, and of very high values of the Mg and K indices.

Figs. 36 and 37 clearly illustrate the biochemical evolutions of insects, as far as hemolymph osmolar effectors are concerned. The great similarity between the body-fluid composition of the apterygote *Petrobius* and the other arthropods is an excellent indication of the fact that primitive insects emerged from the common arthropodial trunk with an internal medium of the "basic" type, *i.e.* with sodium chloride as almost the only osmolar effector. The same type of hemolymph composition has been retained by the modern Palaeoptera, as well as by the three orders originally derived from three distinct stocks of Neoptera exopterygotes (according to Jeannel, 1949): Plecoptera, Dictyoptera and Heteroptera. But, in

TABLE IX

INORGANIC CATIONS IN THE INSECT HEMOLYMPH

(Florkin and Jeuniaux, 1964; where references to literature will be found)

Insect	mequiv./l				Sum of cations	Indices (% of the sum)			
	Na	K	Ca	Mg		Na	K	Ca	Mg
Apterygotes									
Petrobius maritimus	208	5.8	—	—					
Exopterygotes—Palaeoptera									
Ephemeroptera									
Larvae: *Ephemera danica*	103	18	—	—					
Odonata									
Larvae: *Aeschna grandis*	145	9	7.5	7.5	169	85.7	5.3	4.4	4.4
Aeschna cyanea	142	8	—	—					
Aeschna sp.	143	4.3	16	—					
Aeschna sp.	134.7	5.4	7.5	6.0	153.6	87.7	3.5	4.9	3.8
Aeschna sp.	179.3	4.5	20.4	12.3	216.5	82.8	2.1	9.4	5.7
Libellula depressa	178.3	3.8	18.4	12.0	212.5	83.9	1.8	8.7	5.6
Libellula sp.	152.0	—	16.0	4.8					
Libellula sp.	—	—	7.5	—					
Agrion (*Calopteryx*) sp.	158.0	9.0	—	—					
Agrion virgo	140	8	—	—					
Agrion virgo	145	9	—	—					
Enallagma cyathigerum	139	14	—	—					
Adults: *Aeschna cyanea*	120	21	—	—					
Agrion virgo	145	27.5	—	—					

Dictyoptera									
Larvae: *Periplaneta americana*	100	15.4	3.3	—	174.2	90.1	4.3	2.4	3.1
Adults: *Periplaneta americana*	157	7.6	4.2	5.4					
Periplaneta americana	—	—	8.5	22.9					
Periplaneta americana	—	—	17.8	13.5					
Periplaneta australasiae	—	—	19.4	14.8					
Blabera fusca	—	—	20.2	15.7					
Isoptera									
Larvae: *Cryptotermes havilandi*	103	28	—	—					
Zootermopsis angusticollis	—	—	8.6	17.6					
Adults: *Zootermopsis angusticollis*	—	—	16.8	34.8					
Plecoptera									
Larvae: *Perla bipunctata*	127	12	—	—					
Dinocras cephalotes	117	10	—	—					
Cheleutoptera									
Adults: *Carausius morosus*	21.0	25.0	—	—					
Carausius morosus	14.0	16.0	—	—					
Carausius morosus ("serum")	11	18	7	108	144	7.6	12.5	4.8	75
Carausius morosus	8.7	27.5	16.2	145.0	197.4	4.4	13.9	8.2	73.5
Carausius morosus	15	18	15	106	154	9.7	11.6	9.7	68.8
Orthoptera									
Larvae: *Chorthippus parallelus*	72	30	—	—					
Locusta migratoria migratorioides	60.0	12.0	17.2	24.8	114.0	52.6	10.5	14.9	21.9
Schistocerca gregaria	81.3	5.3	17.8	34.6	139.0	58.6	3.8	12.8	24.9
Adults: *Anabrus simplex*	21.9	15.4	3.0	1.4	41.7	52.5	36.9	7.2	3.4
Chortophaga viridifasciata	108.9	3.4	2.8	21	136.1	80	2.5	2.0	15.4
Gryllotalpa gryllotalpa	233.7	7.3	28.0	10.4	279.4	83.6	2.6	10.0	3.7
Gryllotalpa gryllotalpa	174.0	11.0	—	—					

TABLE IX (continued)

Insect	mequiv./l				Sum of cations	Indices (% of the sum)			
	Na	K	Ca	Mg		Na	K	Ca	Mg
Exopterygotes—Polyneoptera (cont.)									
Orthoptera (Adults) (cont.)									
Locusta migratoria migratorioides	67.4	9.0	15.2	27.0	118.6	56.8	7.6	12.8	22.8
Locusta migratoria migratorioides	75	20	—	—					
Locusta migratoria migratorioides	74	15	—	—					
Locusta migratoria migratorioides	102	22	—	—					
Romalea microptera	56.5	17.9	—	—					
Stenobothrus stigmaticus	61.0	62.0	—	—					
Stenopelmatus longispina	—	—	12.1	29.2					
Tettigonia viridissima	83.0	51.0	—	—					
Dermaptera									
Adults: Forficula auricularia	—	—	32.9	—					
Exopterygotes—Paraneoptera									
Homoptera									
Adults: Cinara cilicia	—	—	21.4	30.4					
Jassidae gn. sp.	59	21	—	—					
Heteroptera									
Adults: Gerris najas	142.0	8.0	—	—					
Notonecta kirbyii	—	—	31.0	18.5					
Notonecta obliqua	155	21	—	—					
Corixa punctata	112	31	—	—					
Hesperocorixia larigata	—	—	7.8	3.5					
Rhodnius prolixus	158.0	6.0	—	—					
Rhodnius prolixus	158.0	4.0	—	—					

Triatoma infestans	—	—	40.9	1.5					
Triatoma megista	133.0	5.0	—	—					
Triatoma neotomae	—	—	16.5	1.0					
Triatoma phyllosoma	—	—	13.3	1.2					
Triatoma protracta	—	—	29.5	1.3					
Cinex lectularius	139.0	9.0	—	—					
Oncopeltus fasciatus	—	—	13.9	52.1					
Palomena prasina	22.0	42.0	—	—					
Unknown: stage: *Oncopeltus fasciatus*	39.5	20.5	11	—					
Endopterygotes—Oligoneoptera									
Megaloptera									
Larvae: *Sialis lutaria*	109	5	15	38	167	65.2	3	9	22.7
Planipennia (= Neuroptera)									
Larvae: *Myrmeleon formicarius*	143.5	8.7	12.1	31.3	195.6	73.3	4.4	6.1	16.2
Adults: *Osmylus fulvicephalus*	92	40	—	—					
Mecoptera									
Adults: *Panorpa communis*	94	38	—	—					
Trichoptera									
Larvae: *Anabolia nervosa*	101	17	—	—					
Chaetopteryx villosa	63.9	9	—	—					
Limnephilus stigma	83	14	—	—					
Philopotamus montanus	109	21	—	—					
Phryganea sp.	69	7	—	—					
Phryganea sp.	92.0	6.8	14.4	51.0	164.2	56.0	4.1	8.8	31.1

Bibliography p. 113

TABLE IX (continued)

Insect	mequiv./l				Sum of cations	Indices (% of the sum)			
	Na	K	Ca	Mg		Na	K	Ca	Mg
Endopterygotes—Oligoneoptera (cont.)									
Diptera									
Larvae: Tipula montium	115	7							
Tipula paludosa + oleracea	84.8	8.2	12.3	16.0	121.3	69.9	6.8	10.1	13.2
Dictenidia bimaculata	39.6	3.7	13.8	14.5	71.6	53.3	5.2	19.3	20.2
Chironomus sp.	104.3	2.1	10.5	14.6	131.5	79.3	1.6	8.0	11.1
Chironomus sp.	92.0	8.0	—	—					
Tabanide sp.	151.0	5.0							
Pegomya sp.	26.0	58.0							
Eristalomyia tenax	99.8	8.0	11.8	13.6	133.2	74.9	6.0	8.9	10.2
Gasterophilus intestinalis	206.0	13.0	7.0	38.0	264.0	78.0	4.9	2.7	14.4
Calliphora erythrocephala	148.0	37.0	—	—					
Pupae: Calliphora erythrocephala	139.6	26.1	20.8	34.3	220.8	63.2	11.8	9.4	15.6
Adults: Stomoxys calcitrans	128.0	11.0	—	—					
Eristalis tenax	193.2	20.9	—	—					
Lepidoptera									
Larvae: Cossus cossus	—	15.5	27.7	40.4					
Cossus cossus	18.4	35.4	51.5	48.0	153.3	12.0	23.1	33.6	31.3
Yponomeuta evonymella	3.2	23.3	17.1	29.7	73.3	4.4	31.8	23.3	40.5
Nymphala nymphaeata	40	29	—	—					
Ephestia kuehniella	17.0	60.0							
Ephestia kuehniella	32.6	32.7	41.2	51.1	157.6	20.7	20.8	26.1	32.4
Galleria mellonella	26.5	36.3	24.4	33.5	120.7	22.0	30.1	20.2	27.7
Phalera bucephala	5.9	49.2	34.2	79.8	169.1	3.5	29.1	20.2	47.2
Euproctis chrysorrhoea	17.9	44.5	20.6	87.9	170.9	10.5	26.0	12.1	51.4

Coleoptera

Larvae: Dytiscus sp.

Species									
Larvae: Dytiscus sp.	115	20							
Colymbetes fuscus	127	19							
[...] japonica	20.2	9.5	15.8	38.8	84.3	24.0	11.3	18.7	46.0
Phryganidia californica	—	—	8.5	52.1					
Apamea sordens	—	38.7	17.1	56.8					
Laphygma exigua	—	—	5.4	56.1					
Prodenia praefica	—	—	6.5	64.3					
Phlogophora meticulosa	12.3	34.9	35.5	68.4	151.1	8.1	23.1	23.5	45.3
Prodenia eridania	22.3	39.7	18.4	14.3	94.7	23.5	41.9	19.4	15.1
Peridroma margaritosa	—	—	8.7	85.2					
Estigmene acraea		—	5.1	10.2					
Amathes xanthographa	24.1	29.2	40.4	104.2	197.9	12.2	14.8	20.4	52.6
Triphaena pronuba	16.1	35.6	56.0	70.9	178.6	9.0	19.9	31.4	39.7
Mamestra brassicae	4.3	53.6	17.9	99.2	175.0	2.5	30.6	10.2	56.7
Diataraxia oleracea	13.1	43.1	31.9	78.7	166.8	7.9	25.8	19.1	47.2
Melanchra persicariae	10.3	40.3	19.1	79.0	148.7	6.9	27.1	12.8	53.1
Hypocrita jacobaeae	7.3	34.6	25.0	86.7	153.6	4.8	22.5	16.3	56.4
Spilosoma lutea	3.3	56.2	31.4	38.5	129.4	2.6	43.4	24.3	29.7
Bombyx mori	14.0	35.0	—	—					
Bombyx mori	12.3	35.9							
Bombyx mori: 3rd instar	3.4	41.8	30.6	81.1	156.9	2.2	26.6	19.5	51.7
Bombyx mori: 4th molt	6.0	39.4	15.0	88.0	148.4	4.0	26.5	10.1	59.3
Bombyx mori: 5th instar	14.6	46.1	24.5	102.3	187.5	7.8	24.6	13.1	54.6
Bombyx mori: prenymphs	8.2	59.3	26.5	92.6	186.6	4.4	31.8	14.2	49.6
Antheraea mylitta	1.3	49.7	21.9	37.7	110.6	1.2	44.9	19.8	34.1
Actias selene	4.6	51.3	25.6	59.6	141.1	3.3	36.4	18.1	42.2
Sphinx ligustri	—	34.7	30.5	57.5					
Celerio (Deilephila) euphorbiae	11.0	20.0	36.0	—					
Pieris rapae	—	39.0	41.0	66.6					
Pieris rapae	—	96.4	16.6	92.5					
Pieris brassicae	9.0	19.7	—	—					
Pieris brassicae	5.0	30.0	—	—					
Pieris brassicae	22.0	27.0	—	—					
Aglais (Vanessa) urticae	—	43.0	—	—					
Junonia coenia		—	5.2	29.1					
Papilio machaon	13.6	45.3	33.4	59.8	152.1	8.9	29.8	22.0	39.3

Bibliography p. 113

Bibli

TABLE IX (continued)

Insect	mequiv./l				Sum of cations	Indices (% of the sum)			
	Na	K	Ca	Mg		Na	K	Ca	Mg
Endopterygotes—Oligoneoptera (cont.)									
Hymenoptera									
Larvae: Pteronidea ribesii	1.6	43.4	17.5	60.7	123.2	1.3	35.2	14.2	49.3
Tenthredinide sp.	6.0	55.0	—	—					
Neodiprion sertifer	3	38	—	—					
Vespula germanica	48.0	41.0	—	—					
Vespula germanica	26.0	56.4	18.7	23.6	124.7	20.9	45.2	15.0	18.9
Apis mellifica	10.0	45.0	—	—					
Apis mellifica	10.9	30.5	18.2	20.5	80.1	13.6	38.1	22.7	25.6
Apis mellifica	5.0	24.4	7.5	15.8	52.7	9.5	46.3	14.2	30.0
Pupae: Formica rufa	14.7	50.3	14.9	21.6	101.5	14.5	49.6	14.7	21.2
Vespula germanica	22.8	60.8	11.2	19.0	113.8	20.0	53.4	9.9	16.7
Adults: Vespula pensylvanica	—	—	7.1	1.0					
Vespula germanica	93	18.2	1.8	2.6	115.6	80.4	15.7	1.5	2.2
Vespula germanica	153.5	21.9	2.2	0.5	178.1	86.1	12.3	1.2	0.3
Apis mellifica	47.1	27.1	17.8	1	93	50.6	29.1	19.1	1.2

these primitive insects, we already find some indication of the evolutionary tendencies developed later in the more specialized insects, *i.e.* a slight reduction of the sodium chloride and the incorporation of small organic molecules in the bulk of the hemolymph constituents. This tendency develops considerably in the endopterygotes, for the monophyletic origin of this group suggests that the increasing utilization of free amino acids (and other organic molecules) in replacement of chloride occurred very early in the evolution of endopterygotes, probably prior to the divergence of the "panorpoid complex".

It appears that two different tendencies are to be distinguished during the evolution of the different orders from the "panorpoid complex", one being the conservation of a high amount of inorganic cations, the other (represented by Hymenoptera, Lepidoptera and many Coleoptera) being the strong decrease of inorganic cations in the hemolymph. According to Sutcliffe (1963), this last specialization probably occurred independently on at least two occasions, these three orders being derived independently from the panorpoid line.

Insects are not able to control the concentration of inorganic ions in their hemolymph when placed in a more diluted or concentrated medium. However, osmoregulation takes place to some extent through the modification of the aminoacidemia. This is the case for dragonfly larvae and for *Dytiscus marginalis* adults (Schoffeniels, 1960).

Table IX is an exhaustive recapitulation of the numerous data obtained by different authors. The results are expressed in milliequivalents per liter, and in per cent of the sum of the cations ("indices"). For each order, the data concerning larval, pupal and adult stages are presented separately. It may be seen that the indices vary as follows: Na: from 0.6 to 90.1; K: from 1.6 to 53.4; Ca: from 1.2 to 37.6; and Mg: from 0.3 to 75. The significance of the different types of cationic patterns can be discussed from several points of view.

It must be emphasized that a definite picture of the hemolymph cationic pattern is still difficult to present for each order, owing to the lack of representative data for the different developmental stages. With the exception of a few species, data have been accumulated for only one stage in each order. Table IX shows, for instance, that the cationic hemolymph composition of Homoptera and Hetero-ptera is only known for adults, whereas that of Trichoptera and Lepidoptera is practically only known for larvae or nymphs. This fact seems to have been neglected by many authors, who discussed the systematic or phylogenic significance of the hemolymph cationic composition by comparing animals of different ontogenic positions.

The assumption, made by the authors, that the cationic composi-tion of the hemolymph does not vary significantly during meta-morphosis is based on a few cases, mainly of exopterygotes, in which the hemolymphs of both larval and imaginal stages have approximately the same composition (see Table IX: Odonata: *Aeschna cyanea* and *Agrion virgo*; Dictyoptera: *Periplaneta ameri-cana*; Orthoptera: *Locusta migratoria*). This seems also to be true in the case of two endopterygotes: *Bombyx mori* and *Dytiscus* sp.

However, reexamination of the situation among Hymenoptera (Florkin and Jeuniaux, 1964; see also Table IX) led to the conclu-sion that the cationic composition of the hemolymph is greatly altered during metamorphosis, in this order. In the larval and nymphal hemolymph of bees and wasps, the Na index is indeed only 10 to 20, while that of K ranges from 38 to 53, and that of Mg from 16.7 to 30. In the adults, these proportions are reversed, the Na indices being consistently higher (50–86) and those of K and Mg considerably lower (K: 12–29; Mg: 0.3–2.2).

It is clear that one must be careful before generalizing about the different developmental stages of one insect order on the basis of separate results obtained with representatives of only one or two stages.

From the data in Table IX it is possible to recognize some characteristic patterns, bearing in mind that the sampling is ob-viously scattered, and that ontogenic variations are often ignored.

(1) *Apterygotes*: the only important cation is Na.

(2) *Exopterygotes—Palaeoptera*: in Ephemeroptera and Odonata Na is the most important cation (103–179 mequiv./l), the other

cations being of a very low concentration (less than 30 mequiv./l). This seems to be true for larvae as well as for adults.

(3) *Exopterygotes—Polyneoptera*: with the exception of *Carausius*, Na is also the most important ion, but K, Ca and especially Mg tend to become more concentrated than in Palaeoptera. In one case (*Stenobothrus stigmaticus*) the K concentration is similar to that of Na. The situation seems to be the same in larvae and adults.

Cheleutoptera are characterized by a completely different pattern in which Mg replaces Na almost entirely.

(4) *Exopterygotes—Paraneoptera*: the hemolymph of larvae has not been studied. In adults, the situation is not very different from that found in other exopterygotes, with the exception of *Oncopeltus fasciatus* (Mg: 52.1 mequiv./l) and of *Palomena prasina*, in which the K concentration is twice that of Na.

(5) In those Oligoneoptera: Megaloptera, Neuroptera, Mecoptera, Trichoptera and Diptera, Na is also the main cation (indices 53–79). There seems to be no fundamental difference between the ontogenic stages.

(6) *Coleoptera*: the available data are particularly diverse. It may tentatively be proposed to consider the existence of three groups. In the first group, Adephaga, both larval and adult hemolymphs contain a high proportion of Na (110–165 mequiv./l) and a low proportion of K, Ca and Mg—a pattern similar to that found in Polyneoptera. In a second group, corresponding to the Phytophaga, and in which larval and adult stages of Chrysomelidae, Curculionidae, and Cerambycidae have been recently investigated (Naoumoff and Jeuniaux, unpublished results), the hemolymph of both stages contain a very low amount of Na, while K, Ca and especially Mg are at high concentration. This pattern is similar to that found in Lepidoptera. Finally, a third group may be presumed, in which the adult hemolymph contains more Na, and less K and Mg, than the larval hemolymph (for instance: Scarabaeidae).

(7) *Lepidoptera*: In so far as larval and nymphal hemolymphs are considered, Lepidoptera are characterized by a low proportion of Na (from traces to 30 mequiv./l; indices 1.2–23), higher proportions

of K, of Ca (generally 10–60 mequiv./l) and large amounts of Mg (30–100 mequiv./l; indices 30–50). The spectrum of Na concentration is situated below the lowest limit of the values recorded for animals outside the class of insects (with one exception: *Anodonta*).

The spectrum of Mg concentration can be superimposed on the spectrum found in sea animals, but it is situated above the highest values recorded for fresh-water or terrestrial invertebrates and for vertebrates. This is also the case with potassium. The hemolymph of larval and pupal stages of Lepidoptera thus appears to have a very specialized cationic pattern different from that of other animal phyla.

The cationic pattern of adult hemolymph is only known in the case of two species: *Bombyx mori* and *Telea polyphemus*. These data seem to indicate that adult hemolymph does not differ from that of the larvae or pupae. However, in other species such as *Barathra brassicae* and *Pieris brassicae*, the adult hemolymph appears to be less specialized than the larval hemolymph, the proportion of sodium being doubled in the adults, while that of magnesium is significantly reduced (Naoumoff and Jeuniaux, unpublished).

(8) *Hymenoptera*: In larvae and nymphs of Symphyta and Aculeata, the most important cations are K and Mg. The situation is very different in the adult Aculeata, in which the cationic pattern shows a high Na index (50–80), less K (index 12–30) and only minute amounts of Ca and Mg.

In order to account for the normally functioning, excitable tissues in such insects with a hemolymph rich in K and poor in Na, several authors have postulated that an important proportion of the cations exists not as free ions in the hemolymph, but in a combined form (Barsa, 1954; Bishop *et al.*, 1925; Buck, 1953; Clark and Craig, 1953). However, this is not the case for *Antherea polyphemus*, in the hemolymph of which no evidence has been detected for any binding of K, while only 15–20% of the Ca and Mg is bound to macromolecules (Carrington and Tenney, 1959).

For Boné (1944) and for Tobias (1948a) the explanation of the different types of cationic patterns is dietetic. In their opinion, zoophagous insects would tend to have high Na, and phytophagous insects high K and Mg in their hemolymph. This relationship appears clearly in many cases, but some insects (grasshoppers, *Tipula* larvae, *Hydrophilus* adults, *Geotrupes*, etc.) contradict this statement, as Boné himself pointed out.

Insects, being mainly terrestrial and therefore unable to absorb cations from a fluid habitat, can only rely on food to insure the steady state of the concentration of cations in their hemolymph, which is the result of the equilibrium between ingestion and excretion. It is therefore reasonable to compare the concentrations of these cations, per 1000 g of fresh food, or per 1000 ml of hemolymph. Table X shows that when the insects considered are phytophagous, the specialized pattern of cations in hemolymph is always due to the dilution of potassium and calcium, and to the concentration of magnesium. With respect to sodium, we can see that either dilution or concentration takes place. The non-phytophagous insects, with the specialized pattern, which appear in Table X are the bee larvae, eating honey, the larva *Cossus cossus*, which eats wood, and *Galleria mellonella* which feeds on the wax comb in the beehive. The table shows that the bee larva concentrates all the cations of honey, while *Cossus* and *Galleria* dilute the potassium, the magnesium and the calcium of their food and concentrate the sodium.

From this survey, it can be seen that the concept according to which some insects have a high potassium and a low sodium content as a consequence of eating foliar food, and others a high sodium and a low potassium content because they do not consume this kind of food, is not acceptable.

Duchâteau *et al.* (1953) proposed a hypothesis involving both phylogenic and dietetic considerations in order to explain the diversity of cationic patterns. According to the classical view of insect taxonomy, the cationic pattern of Palaeoptera (high-sodium type) is considered a primitive pattern among insects, not dissimilar from that of other animal taxa and of apterygotes, if we consider the "indices" of each cation (see Table IX).

The pattern found in other insect orders, especially in Lepidoptera, is strikingly different from the type defined above, and

TABLE X

COMPARISON BETWEEN THE CATIONIC COMPOSITION OF THE HEMOLYMPH OF INSECTS

(Florkin and Jeuniaux, 1964)

Food and organism	mequiv./kg fresh food or /l hemolymph					Indices			
	Na	K	Ca	Mg	Σ Cations	Na	K	Ca	Mg
Lettuces (Lettuca sativa)[a]	13.0	87.2	—	—					
Periplaneta americana, adults[a]	113.0	25.6	—	—					
Romalea microptera, adults[a]	56.5	17.9	—	—					
Ivy, leaves (Hedera helix)[f]	35.9	147.6	665.0	53.1	900.6	4.0	16.4	73.8	5.8
Privet, leaves (Ligustrum vulgare)[f]	46.4	152.1	824.5	39.9	1062.9	4.4	14.3	77.6	3.8
Carausius morosus, adults[f]	8.7	27.5	16.2	145.0	197.4	4.4	13.9	8.2	73.5
Horse blood, total[b]	84.8	31.4	1.7	3.3	121.4	69.9	25.9	1.6	2.7
Gasterophilus intestinalis, larvae[c]	206.0	13.0	7.0	38.0	264.0	78.0	4.9	2.7	14.4
Poplar, wood (Populus sp.)[f]	16.0	126.0	1471.0	113.9	1726.9	0.9	7.3	85.2	6.6
Cossus cossus, larvae[f]	18.4	35.4	51.5	48.0	153.3	12.0	23.1	33.6	31.3
Wax[f]	12.8	347.2	257.0	215.3	832.4	1.5	41.7	30.9	25.9
Galleria mellonella, larvae[f]	26.5	36.3	24.4	33.5	120.7	22.0	30.1	20.2	27.7
Mulberry tree, leaves[e]	11.3	59.0	—	—					
Bombyx mori, larvae[e]	12.2	35.9	—	—					
Carrot, leaves (Daucus carota)[f]	25.6	176.9	214.5	35.6	422.6	5.7	39.1	47.4	7.9
Papilio machaon, larvae[f]	13.6	45.3	33.4	59.8	152.1	8.9	29.8	22.0	39.3

Potato, leaves (*Solanum tuberosum*)[f]	tr	144.5	128.6	85.9	359.0	—	40.3	35.8	23.9
Leptinotarsa decemlineata, adults[f]	3.5	65.1	47.5	188.3	304.4	1.1	21.4	15.6	61.9
Leptinotarsa decemlineata, adults[f]	2.0	54.9	43.4	146.9	247.2	0.8	22.2	17.6	59.4
Currant-bush, leaves (*Ribes grossulariae*)[f]	tr	249.1	271.2	53.6	573.9	—	43.4	47.3	9.3
Pteronidea ribesii, larvae[f]	1.6	43.4	17.5	60.7	123.2	1.3	35.2	14.2	49.3
Honey[a]	4.7	13.1	2.7	1.8	22.3	21.1	58.7	12.1	8.1
Apis mellifica, larvae[f]	10.9	30.5	18.2	20.5	80.1	13.6	38.1	22.7	25.6

[a] Tobias (1948a). [b] Aberhalden (1898). [c] Levenbook (1950). [d] McCance and Widdowson (1946). [e] Tobias (1948b). [f] Duchâteau *et al.* (1953).

Bibliography p. 113

appears to be a special evolutionary development, found also in other advanced groups, such as certain Coleoptera and in the larval stages of Hymenoptera. This specialized type can be regarded as a systematic characteristic, linked to the genotype controlling the synthesis of the enzymes playing a role in the regulation. We can take into consideration the notion of the evolution of Lepidoptera, Coleoptera and Hymenoptera parallel to the evolution of the angiosperms, and suggest that the speciation along this phylogenic line has been accompanied by a kind of regulation of the steady state of the cationic concentrations in the hemolymph, leading to a low sodium, a high potassium and a high magnesium pattern.

When the insects of these specialized groups adapt themselves secondarily to another form of food, as for example in the case of the wasp and bee larvae, of *Cossus* and of *Galleria*, this ecological change suggests the acquisition of new regulatory processes, maintaining the specialized pattern.

On the other hand, it is true that insects belonging to the orders which have not acquired the specialized type, can very well adopt phytophagous habits without acquiring the pattern of cationic concentrations which is found in Lepidoptera and Hymenoptera. Clearly, this pattern is not a question of food, it is a question of taxonomy.

The muscles of *Carausius morosus* and of Lepidoptera larvae function well and show action potentials in salines of a composition reproducing the cationic pattern of their hemolymph. This points to the fact that the mechanism of neuromuscular transmission must be of such a nature as to allow the muscle function to operate in media containing a high concentration of potassium, an extremely high concentration of magnesium and almost no sodium. Hoyle (1954) suggests that mechanisms similar to those of crustaceans could be adapted to function in such media, while the vertebrate mechanism could not be so adapted. Hoyle suggests that the type of cationic pattern of the "specialized" insects may be a way of reducing spontaneous activity and speed of movement. For instance, the level of potassium in phytophagous insects is reduced by fasting, and it has been suggested that effects of this kind may be at work in building up the hypertensive excited state of migratory locusts (Ellis and Hoyle, 1954; Hoyle, 1954).

It appears that insects have on several occasions developed a regulation of the inorganic constituents of hemolymph in which the cationic pattern is not compatible with the function of the nerves and muscles of species belonging to other categories of insects or other animals.

This specialization appears, as already pointed out, to be linked with speciation parallel with the development of angiosperms. The ecological interest of the acquisition of the specialized hemolymph type may perhaps be linked with a behavioral aspect of relative inactivity, maintaining the larval stages in the midst of abundant foods, as is the case with caterpillars.

From this point of view, it is particularly interesting to note the striking modification of the ratio Na/K during the metamorphosis of bees and wasps, starting from the resting larvae, with the specialized type of cationic pattern, to the exceedingly active adults, with a hemolymph containing large amounts of Na.

From the point of view of phylogeny, it seems therefore that the adaptation to an entirely vegetable diet, and to a sedentary life in the midst of food, has been developed independently in different orders, and generally as a particular feature of the larval stage. The adult stage generally retains the basic and primitive cationic pattern. According to their phyletic position and to the specialized pattern of both larval and adult hemolymphs, the Coleoptera of the families Curculionidae, Cerambycidae and Chrysomelidae, and probably also the Lepidoptera are, among insects, the most fully adapted to phytophagous habits.

The variability of blood volume is a very important physiological adaptation in insects. They have no use for oncosmotic pressure as insured by serum albumin or serum globulins in vertebrates; nor do they have need of the aggregation of amino acids into the proteins at work in the vertebrate plasma. On the other hand, they regulate the osmolar concentration of the blood in spite of plasma volume changes. These two conditions are met by the existence of free amino acids in solution. These afford the nitrogen pressure insuring the delivery of amino acids to cells, and their concentrations exert a regulation of osmotic pressure in cases of blood-volume changes. The amino acid pool of the hemolymph is replenished not only through the amino acids provided with food, but also, in certain cases, from tissue sources. In crustacea, growth is accompanied by molting, but in insects this aspect, which is also at work, is complicated by the succession of forms of different ecologically adapted characteristics, and metamorphosis implies a transfer of cell constituents for the rapid formation of the imago at the end of the pupal stage, for example. In a number of insects, the composition of the blood is, contrary to what obtains in vertebrates, very similar to that of cells. This applies, for instance, as we have seen, to Lepidoptera, Hymenoptera and certain specialized Coleoptera. This pattern is well suited to an internal medium which is rapidly invaded by the products of cell lysis and which is rapidly tapped for the formation of new tissues.

BIBLIOGRAPHY

ABDERHALDEN, E., Zur quantitativen vergleichenden Analyse des Blutes, *Z. Physiol. Chem.*, 25 (1898) 65–115.

BALDWIN, E., *An Introduction to Comparative Biochemistry*, 3rd ed., Cambridge University Press, Cambridge, 1948.

BALDWIN, E., *The Nature of Biochemistry*, Cambridge University Press, Cambridge, 1962.

BARSA, M. C., The behaviour of isolated hearts of the grasshopper *Chortophaga viridifasciata*, and the moth, *Samia walkeri* in solutions with different concentrations of sodium, potassium, calcium and magnesium, *J. Gen. Physiol.*, 38 (1955) 79–92.

BISHOP, G. H., A. P. BRIGGS AND E. RONZONI, Body fluids of the honey bee larva. II. Chemical constituents of the blood and their osmotic effects, *J. Biol. Chem.*, 66 (1925) 77–88.

BONÉ, G. J., Le rapport sodium/potassium dans le liquide coelomique des insectes. I. Ses relations avec le régime alimentaire, *Ann. Soc. Roy. Zool. Belg.*, 75 (1944) 123–132.

BUCK, J. B., Physical properties and chemical composition of insect blood, in K. D. ROEDER (Ed.), *Insect Physiology*, Wiley, New York, 1953, pp. 147–190.

CARRINGTON, C. B., AND S. M. TENNEY, Chemical constituents of haemolymph and tissue in *Telea polyphemus* Cram. with particular reference to the question of ion binding, *J. Insect Physiol.*, 3 (1959) 402–413.

CLARK, E. W., AND R. CRAIG, The calcium and magnesium content in the hemolymph of certain insects, *Physiol. Zool.*, 26 (1953) 101–107.

DUCHÂTEAU, G., M. FLORKIN AND J. LECLERCQ, Concentrations des bases fixes et types de composition de la base totale de l'hémolymphe des insectes, *Arch. Intern. Physiol. Biochim.*, 61 (1953) 518–549.

ELLIS, P. E., AND G. HOYLE, A physiological interpretation of the marching of hoppers of the African migratory locust (*Locusta migratoria migratorioides* R. and F.), *J. Exptl. Biol.*, 31 (1954) 271–279.

FLORKIN, M., AND CH. JEUNIAUX, Hemolymph composition, in M. ROCKSTEIN (Ed.), *Physiology of Insecta*, Vol. 3, Academic Press, New York, 1964, pp. 109–152.

HOYLE, G., Changes in the blood potassium concentration of the African migratory locust (*Locusta migratoria migratorioides* R. and F.) during food deprivation and the effect on neuromuscular activity, *J. Exptl. Biol.*, 31 (1954) 260–270.

JEANNEL, R., Les insectes, in P. P. GRASSÉ (Ed.), *Traité de Zoologie*, Vol. 9, Masson, Paris, 1949, pp. 1–17.

LEVENBOOK, L., The composition of horse bot fly (*Gastrophilus intestinalis*) larva blood, *Biochem. J.*, 47 (1950) 336–346.

LOCKWOOD, A. P. M., AND P. C. CROGHAN, Composition of the haemolymph of *Petrobius maritimus* Leach, *Nature*, 184 (1959) 370–371.

MCCANCE, R. A., AND E. M. WIDDOWSON, *The Chemical Composition of Foods*, 2nd ed., H. M. Stationery Office, London, 1946.

SCHOFFENIELS, E., Rôle des acides aminés dans la régulation de la pression osmotique du milieu intérieur des insectes aquatiques, *Arch. Intern. Physiol. Biochim.*, 68 (1960) 507–508.

SUTCLIFFE, D. W., The composition of haemolymph in aquatic insects, *J. Exptl. Biol.*, 39 (1962) 325–344.

SUTCLIFFE, D. W., The chemical composition of haemolymph in insects and some other Arthropods, in relation to their phylogeny, *Comp. Biochem. Physiol.*, 9 (1963) 121–135.

TOBIAS, J. M., Potassium, sodium and water interchange in irritable tissues and haemolymph of an omnivorous insect *Periplaneta americana*, *J. Cellular Comp. Physiol.*, 31 (1948a) 125–142.

TOBIAS, J. M., The high potassium and low sodium in the body fluid and tissues of a phytophagous insect, the silkworm *Bombyx mori* and the change before pupation, *J. Cellular Comp. Physiol.*, 31 (1948b) 143–148.

Chapter 8

Definition, in Terms of Molecular Evolution, of Special Characteristics in the Carbohydrate Metabolism of Insects

A number of adaptations of insects to different habitats by way of stages traversed during metamorphosis are linked to the nature of the carbohydrates of insect hemolymph.

It has been known for a long time that insect hemolymph generally contains small amounts of fermentable sugars, almost no saccharose, and little if any glycogen. The reducing power of the hemolymph is sometimes relatively high, but the greater part of this reducing power is due to substances non-saccharidic in nature, such as ascorbic acid, α-ketonic acids, uric acid, tyrosine and other phenols, and doubtless also to many other unknown substances.

The explanation of such an unusually low concentration of fermentable sugars in an internal medium resides in the discovery by Wyatt and Kalf (1956, 1957) of the existence in the hemolymph of a non-reducing dimer of α-glucose, trehalose, in high concentration. Hemolymph trehalose appears to be a form of carbohydrate transport, peculiar to the class of insects.

The data concerning the amount of substances fermentable by yeast are presented in Table XI. The nature of these substances has been determined in only a few cases. In the adult bee, the fermentable substances are fructose (30–40%) and glucose (60–80%) (Von Czarnovsky, 1954). Fructose is also present in rather large amounts in the hemolymph of *Gasterophilus intestinalis* (Levenbook, 1947, 1950) and glucose in that of *Phormia regina*, in which its concentration increases in the adult stage (Evans and Dethier, 1957). These high levels of fructose and glucose appear, however, to be exceptional in the hemolymph of insects.

The concentration of trehalose in a number of representative insects is shown in Table XI. Trehalose is generally present in large

TABLE XI

CONCENTRATION OF TOTAL FERMENTABLE SUGARS (EXPRESSED IN GLUCOSE, (mg/100 ml), AND OF GLUCOSE, FRUCTOSE AND TREHALOSE IN THE HEMOLYMPH OF INSECTS (mg/100 ml)

(Florkin and Jeuniaux, 1964)

Species	Stage	Fermentable sugars (as glucose)	True glucose	Fructose	Trehalose
Dictyoptera					
Periplaneta americana[b]	?	30	—	—	—
Leucophaea maderae[m]	?	65	—	—	580–780
Orthoptera					
Schistocerca gregaria	Larvae	—	—	traces[i]	800–1500[h]
Coleoptera					
Hydrophilus piceus	Adults	5–31[e]	—	—	300–500[b]
Popillia japonica	Larvae	69[k]	—	—	—
Chalcophora mariana	Larvae	—	—	—	4800–5300[b]
Erbates faber	Larvae	—	—	—	3200[b]
Hymenoptera					
Diprion hercyniae[o]	Larvae	—	28	—	926
Apis mellifica	Adults	1000–4000[n]	600–3200[n]	200–1600[n]	600–1200[b]
Lepidoptera					
Phalera bucephala	Larvae	40[f]	—	—	—
Prodenia eridania	Larvae	11[a]	—	—	—
Bombyx mori	Larvae	9–28[d,e]	1–3[p]	1–2[p]	400–500[p]
	Nymphs	18–50[d,e]	3–5[p]	1–2[p]	202[p]
	Adults	16[d,e]	—	—	—
Deilephila euphorbiae	Larvae	traces[g]	—	—	—
Deilephila elpenor	Nymphs	—	—	—	800–1900[b]
Galleria mellonella	Larvae	—	21[o]	—	1700[o]
Hyalophora cecropia[o]	Larvae	—	—	—	1200
Hyalophora cecropia	Nymphs	—	0–8	—	400–600
	Adults	—	—	—	650–1150
Telea polyphemus[o]	Larvae	—	—	—	1306
Diptera					
Gasterophilus intestinalis[i,j]	Larvae	95	10	184–294	—
Calliphora erythrocephala[i]	Larvae	—	—	traces	—
Phormia regina[c]	Larvae	—	70–125	—	absent
	Adults	—	up to 600	—	598

[a] Babers, 1938; [b] Duchâteau and Florkin, 1959; [c] Evans and Dethier, 1957; [d] Florkin, 1936; [e] Florkin, 1937; [f] Hemmingsen, 1924; [g] Heller and Moklowska, 1930; [h] Howden and Kilby, 1956; [i] Levenbook, 1947; [j] Levenbook, 1950; [k] Ludwig, 1951; [l] Todd, 1957; [m] Todd, 1958; [n] Von Czarnowsky, 1954; [o] Wyatt and Kalf, 1957; [p] Wyatt et al., 1956.

amounts in the hemolymph of all the insects studied so far, with the remarkable exception of the larva of *Phormia regina* (Evans and Dethier, 1957).

In vertebrates, the cells have very little glucose, but this fermentable sugar is the form in which the blood carries carbohydrate cellular food. Blood glucose is liberated by the liver cells from glycogen, owing to the presence of glucose-6-phosphatase in these cells. The glycogen itself is derived from a variety of substances: glucose, fructose, galactose, including some non-glucidic sources: amino acids, pyruvate, lactate, etc. (gluconeogenesis). This is due to the presence of fructose-1,6-diphosphatase, which allows for glucose 1-phosphate and glycogen biosynthesis by a reversal of glycolysis. In vertebrates, therefore, the circulating carbohydrate is glucose and it is largely of endogenous origin, the product of gluconeogenesis (formation of carbohydrates from non-carbohydrate precursors). From there the glucose goes to the cells, crossing the membrane as hexose 6-phosphate. The whole system is controlled by a complicated hormonal system, including the specialized biosynthesis and action of insulin, glucagon and adrenaline.

These systems are as foreign to insects as, for example, is the possession of a cord terminating in a cerebral enlargement contained in a cranial cavity formed by the expansion of skeletal vertebrae. The cells of insects do indeed use glucose, but the liberation of this glucose is carried out inside the cells by the action of the enzyme trehalase. The circulating form is trehalose. It is exceptional for insect hemolymph to contain greater concentrations of monosaccharides than of trehalose (*Phormia regina*: Evans and Dethier, 1957; adult bee: Duchâteau and Florkin, 1959).

The notion of trehalose as the circulating carbohydrate of insects is based on the following observations: (*a*) the cells of many tissues contain an active trehalase: muscles, middle gut, salivary glands (Kalf and Rieder, 1958; Howden and Kilby, 1956; Zebe and McShan, 1959; Duchâteau-Bosson *et al.*, 1963); (*b*) in the course of muscular activity (sustained flight, for example) the concentration of plasma trehalose is lowered: the muscles therefore utilize this trehalose (Evans and Dethier, 1957; Clegg and Evans, 1961; Bücher and Klingenberg, 1958); (c) in the nymphs of *Deilephila elpenor*, just after the termination of diapause (formation of adult tissues) the trehalosemia diminishes (Duchâteau and Florkin, 1959).

It is therefore clear that many tissues utilize, in their metabolism, the trehalose of hemolymph, which is hydrolyzed in the presence of intracellular trehalase.

As we have said, many insect tissues are equipped with the enzyme trehalase, which liberates glucose from the trehalose absorbed from hemolymph. But both epiderm and silk gland appear to be devoid of trehalase, even in the periods of full activity of these tissues, such as during molting and spinning (Zebe and McShan, 1959; Duchâteau *et al.*, 1963). If this is confirmed, it appears that in the specialized tissues of the epiderm and of the silk gland, trehalose is not the source of energy utilized by the tissue. The experiments pursued in the author's laboratory and in the laboratory of T. Fukuda in Japan have shown that the silk glands are able to absorb certain amino acids from the surrounding hemolymph and to incorporate them in silk fibroin. They are also able to metabolize them into pyruvate. They can utilize pyruvate and formate for the synthesis of a number of amino acids (serine, glycine, alanine,

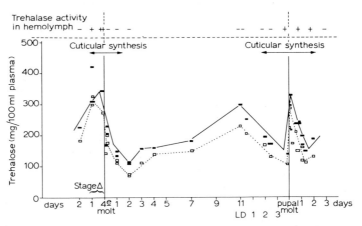

Fig. 38. Modification of trehalose concentration and of trehalase activity in the hemolymph of *Bombyx mori*. ■, anthrone reactive material, in mg trehalose/ml; □, trehalose, mg/100 ml. LD = last defecation. (Duchâteau *et al.*, 1963)

threonine: Bricteux et al.. 1959 and Bricteux-Grégoire et al., 1959. The conclusion drawn is that the silk glands rely for plastic and energy sources on all kinds of organic molecules able to give pyruvate in the metabolism, apparently with the exception of trehalose. With respect to molting, and in the absence of trehalase in the epidermal tissue, the answer is provided by a study, during the ontogenesis of the silkworm, of the variations of trehalosemia during the molt and intermolt periods and by a study of the variations of the trehalase activity in the hemolymph.

The epidermal cells use glucose liberated by the enzymatic hydrolysis of trehalose—a hydrolysis performed not inside the cells, but outside in the hemolymph. The supply of glucose from the trehalose of the hemolymph has been investigated by Duchâteau et al. (1963); this mechanism is illustrated in Fig. 38, which shows the variations in Bombyx mori during the end of larval and the beginning of nymphal life.

The amount of blood trehalose sharply decreases at each molt, and also during the fasting period corresponding to spinning. The fall of blood trehalose during the molts corresponds to the increase of glucose (Florkin, 1937b) observed at the same period. It is related to the release, probably of hormonal nature, of the inhibition of the trehalase present in an inactive state in the hemolymph. In the fat body, an inverse relationship exists between glycogen and trehalose, the former disappearing almost completely at each molt, while the amount of trehalose tends to remain at nearly constant level (Duchâteau et al., 1963).

Trehalose is not introduced into the insect body with the food; it is synthesized by the insect. Treherne (1958a and b) has shown that glucose or fructose marked with an atom of radioactive carbon are rapidly converted into marked trehalose, whether the monosaccharides are introduced per os or by injection into the hemolymph. Candy and Kilby (1961) have studied the mechanism of trehalose biosynthesis from glucose. Glucose is converted into glucose 6-phosphate. This is coupled with a molecule of glucose, with uridine diphosphate–glucose (UDPG) functioning as a glucose donor.

The synthesis of trehalose phosphate from glucose and UDPG is catalyzed by a specific enzyme: trehalose-phosphate–UDP glucosyl-transferase. Trehalose phosphate is hydrolyzed into trehalose + phosphate under the action of trehalose phosphatase. This biosynthetic pathway of trehalose appears to be a lateral extension of the glycolysis chain or of the gluconeogenesis chain, this extension being attached at the level of G-6-P. This biosynthetic pathway has been shown to be active in the fat body of insects, and this fat body has been considered to be the principal, if not the only, location of the biosynthesis of trehalose. The fact that the trehalose of hemolymph may, at least in certain circumstances, be derived from the glycogen of the fat body is proved by a number of observations. At each period of intense activity or of fasting, when trehalosemia is lowered, the content of glycogen in the fat body also diminishes and in much greater proportions (Saito, 1963; Duchâteau et al., 1963). One of the arguments in favor of the metabolic relationships between fat-body glycogen and hemolymph trehalose is to be found in the observations of Steele (1963) on *Periplaneta americana*, in which an injection of corpora allata results in a rapid rise in trehalosemia, with a simultaneous fall in fat-body glycogen. According to Steele, this experiment indicates the liberation by the corpora allata of a hyperglycemic hormone, acting as an activator of glycogen phosphorylase, and catalyzing the transformation of glycogen into glucose 1-phosphate (G-1-P).

All these observations favor the notion that fat-body glycogen is at least one of the sources for the production of trehalose, and probably the only one during fasting.

Recent experiments by Bricteux-Grégoire et al. (1964, 1965) suggest that the fat body is not the only site of trehalose synthesis and that the pathway of this biosynthesis does not necessarily pass through the stage of glycogen. In more or less undernourished silkworms, the radioactivity of labelled pyruvate, injected into the hemolymph is found, four hours after the injection, only in the hemolymph trehalose, and not in the fat-body trehalose. If labelled G-1-P is injected, it is also only at the level of hemolymph trehalose that the most intense incorporation is found. The glycogen of fat body is likewise radioactive four hours after the injection of labelled G-1-P, but glycogen synthesis is twice as important in normally fed silkworms as in underfed silkworms. These results show that the amount of food consumed plays a role in the determination of the biosynthetic pathway followed by the materials of gluconeogenesis. How can this happen? It could result from a mechanism of hormonal regulation of the type described by Steele (1963, see above). But it also seems possible that the level of trehalosemia can control the synthesis of trehalose and of glycogen. Above certain concentration limits, trehalose inhibits trehalose phosphate–UDP glucosyltransferase, so that the resulting increase of UDPG concentration would activate glycogen synthesis (Murphy and Wyatt, 1964). In the very active phenomena of sustained flight in some insects, for instance, Diptera and Hymenoptera, carbohydrate is the only fuel used and is supplied to the muscles in the form of trehalose, which appears to represent an immediately available reserve which is used for a very marked increase in respiration when the animal attains full flight.

This appears to be the result of the acquisition of another lateral extension to the glycolysis chain (see Chapter 4) in which the reduced NAD generated in the usual system is used to reduce dihydroxyacetone phosphate by the action of an NAD-dependent cytoplasmic α-glycerophosphate dehydrogenase. The α-glycerophosphate diffuses into the mitochondrial department, where mitochondrial-bound α-glycerophosphate dehydrogenase is located, and enters directly into the respiratory chain. The appearance, in insects, of an unusual end-product of glycolysis, α-glycerophosphate, which is linked with the adoption of a certain mode of rapid flight in bees and flies, is due to the relative lack of lactic dehydrogenase and to a great concentration of α-glycerophosphate dehydrogenase. The fact that the lactic dehydrogenase is relatively deficient should lead to an accumulation of reduced NAD and, consequently, to a deficit in the NAD necessary for the oxidation of glycerol 3-phosphate. But the increased concentration of α-glycerophosphate dehydrogenase prevents this effect. It reduces dihydroxyacetone phosphate to α-glycerophosphate, while oxidizing reduced NAD to NAD. The result is an accumulation of α-glycerophosphate, which accounts not only for the α-glycerophosphate cycle active in sustained flight of bees and flies, but also leads to another adaptive radiation of ecological importance: the diapause of nymphs (Schneiderman and Williams, 1953) or of eggs.

Andrewartha (1952) has reviewed the biological aspects of the insect adaptation called diapause. Most commonly, and in insects adapted to the cool temperate zone of the Northern Hemisphere, diapause occurs in the stage of metamorphosis which coincides with the lower temperatures obtaining during the succession of seasonal changes. During the nymphal diapause undergone by a number of Lepidoptera, the respiration intake falls very markedly: to 1/50 of the prediapausal level, and no longer responds to carbon monoxide, cyanide, and azide—all substances which inhibit cytochrome oxidase. This latter effect is due to a lowering of the cytochrome c concentration, from which results a high excess of cytochrome oxidase relative to cytochrome c. In this situation, the low level of cytochrome c causes a fresh formation of reduced NAD, which is reoxidized by an α-glycerophosphate dehydrogenase, the resulting α-glycerophosphate being dephosphorylated into glycerol in the tissues. This glycerol protects the insect tissues from the low-temperature effects. Salt (1959) has shown that in the sawfly *Bracon cephi* the concentration of glycerol in the larva reaches values as high as 25% of fresh tissues. In these conditions, the larva withstands temperatures as low as $-40°$ C. In the nymphal diapause of *Hyalophora (Samia) cecropia*, glycerol may reach a concentration of 3.5% in the hemolymph. The glycerol comes from glycogen through trehalose. It appears on the day of nymphosis and increases until it reaches a maximum after one or two months (Wyatt and Meyer, 1959; Wilhelm *et al.*, 1961).

In another saturniid, *Philosamia (Samia) cynthia*, the diapausing nymphs accumulate less glycerol, but if exposed to lower temperatures, the glycerol increases in the hemolymph (Wilhelm, 1960). This is not the only biochemical change which is observed in the blood of hibernating nymphs. Data on free amino acids in larvae and nymphs in diapause are available for *Sphinx ligustri* (Figs. 39

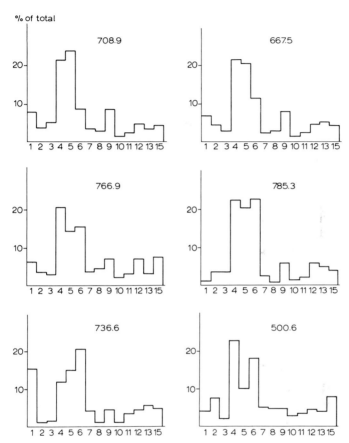

Fig. 39. *Sphinx ligustri*, caterpillars. Figures along the ordinates represent percentages of the sum of the 14 free amino acids considered, the total of which is indicated in each profile (in mg/100 ml plasma). The figures along the abscissae correspond to the different amino acids indicated in Table XII. (Duchâteau and Florkin, 1955)

and 40) and *Smerinthus ocellatus* (Table XII). Alanine levels appear to be much higher in these diapausing nymphs than in their corresponding larvae—another aspect of protection against the effects of low temperature. This is not the case with the non-hibernating nymphs of *Euproctis chrysorrhoea* (Table XII) nor, except for a short time following nymphosis, with another non-hibernating

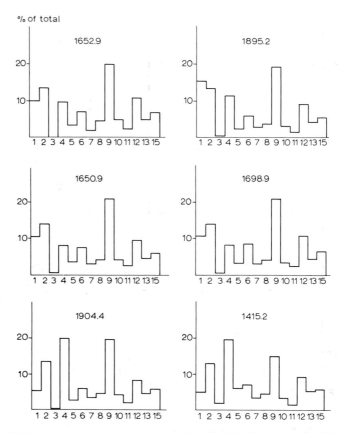

Fig. 40. *Sphinx ligustri*, nymphs. See legend to Fig. 39. (Duchâteau and Florkin, 1955)

nymph, *Bombyx mori*. If we compare the series of data on amino-acidemia in nymphs and in caterpillars of *Sphinx ligustri*, we come to the conclusion that there is less variability of pattern among the diapausing nymphs than among the caterpillars.

The brilliant work of C. M. Williams (1951–1952) has focussed attention on *Hyalophora cecropia*, in which a reaction of the organism to the external conditions is found when exposure to a low temperature, exerting an influence on the control of the suspension of nymphal diapause, is followed by exposure to a higher temper-

TABLE XII

COMPARISON OF CATERPILLAR AND PUPA IN LEPIDOPTERA; CONCENTRATION
OF APPARENT AMINO ACIDS (mg/100 ml OF HEMOLYMPH PLASMA)

(Duchâteau and Florkin, 1958)

Amino acids	Euproctis chrysorrhoea L.				Smerinthus ocellatus	
	Caterpillars (reared on pear leaves)		Pupae		Cater-pillars	Pupae ♀ (hiber-nation)
	May 1952	June 1954	June 1952	June 1954	Oct. 1956	April 1955
1. Alanine	33.0	—	37.9	—	27.7	171.0
2. Arginine	44.8	58.7	99.2	114.7	19.4	319.9
3. Aspartic acid	9.0	22.2	24.2	33.7	27.7	1.2
4. Glutamic acid	303.2	343.6	279.0	350.5	202.2	127.1
5. Glycine	94.3	48.9	63.7	46.3	52.6	38.1
6. Histidine	107.5	161.8	60.5	49.5	83.1	63.6
7. Isoleucine	15.1	32.9	30.6	45.3	12.5	54.2
8. Leucine	13.3	23.6	35.5	43.7	8.3	49.2
9. Lysine	50.5	105.3	225.8	103.7	77.6	315.7
10. Methionine	1.8	13.8	0.0	21.1	8.3	78.8
11. Phenylalanine	8.1	15.1	18.5	11.1	9.7	23.7
12. Proline	129.7	157.8	101.6	124.3	23.5	241.5
13. Threonine	30.8	54.7	38.7	47.4	34.6	55.1
14. Tyrosine	0.0	5.3	5.6	20.0	30.1	16.1
15. Valine	29.8	49.3	46.0	63.7	83.5	89.8
Total	870.9		1066.8		700.8	1645.0

ature. When diapausing nymphs are kept at different temperatures, a consistent result is observed not only for *Deilephila euphorbiae* (Table XIII) but also for *Saturnia pavonia* (Table XIV) and *Sphinx ligustri* (Table XV) (Bricteux-Grégoire *et al.*, 1960). At a low temperature, the concentration of alanine rises in the hemolymph and the concentration of total glutamic acid falls; the reverse is

Bibliography p. 130

TABLE XIII

Deilephila euphorbiae, DIAPAUSING NYMPHS; CONCENTRATION OF APPARENT AMINO ACIDS

(mg/100 ml OF HEMOLYMPH PLASMA)

(Bricteux-Grégoire *et al.*, 1960)

	From 19-11-56 to 9-1-57				From 19-11-56 to 27-2-57			
	25°	14°	5°	2°	25°	14°	5°	2°
Alanine	54.4	116.9	185.5	175.8	24.3	70.1	161.0	272.8
Arginine	136.9	110.0	122.0	116.9	120.6	199.3	164.8	196.0
Aspartic acid	13.3	11.7	9.9	13.1	16.4	19.0	16.2	16.7
Glutamic acid	273.9	120.2	167.4	44.2	242.5	164.2	106.1	54.0
Glycine	41.8	42.2	37.9	52.3	38.1	48.5	61.8	51.7
Histidine	129.5	96.0	82.4	89.9	141.0	121.6	95.5	95.0
Isoleucine	67.4	77.7	90.7	80.1	71.6	71.6	61.2	62.3
Leucine	76.2	91.6	104.7	92.3	79.5	79.1	70.5	72.9
Lysine	311.6	275.7	253.9	248.6	341.8	305.2	259.0	234.8
Methionine	52.6	36.7	46.2	42.5	40.3	53.0	41.8	50.2
Phenylalanine	34.0	32.3	30.5	26.2	35.1	32.1	21.8	23.6
Proline	188.0	226.5	160.8	175.0	201.5	225.4	166.9	111.7
Threonine	83.6	79.9	70.9	59.7	95.5	76.9	63.7	70.7
Tyrosine	31.8	88.7	84.1	100.6	52.2	37.3	78.0	68.5
Valine	95.5	92.4	106.3	98.1	98.5	99.3	82.4	79.8
Total	1590.5	1498.5	1553.2	1415.3	1598.9	1602.6	1450.7	1460.7

TABLE XIV

Saturnia pavonia, DIAPAUSING NYMPHS; CONCENTRATION OF APPARENT
AMINO ACIDS (mg/100 ml OF HEMOLYMPH PLASMA)

(Bricteux-Grégoire *et al.*, 1960)

	From 19-11-56 to 9-1-57			
	$25°$	$14°$	$5°$	$2°$
Alanine	42.7	102.9	396.0	368.2
Arginine	136.2	137.2	103.7	137.6
Aspartic acid	—	25.1	38.0	31.6
Glutamic acid	513.0	366.8	242.5	163.6
Glycine	—	79.2	67.2	59.5
Histidine	108.3	98.9	70.9	94.8
Isoleucine	—	66.0	73.1	80.0
Leucine	—	92.3	92.1	104.1
Lysine	—	302.1	254.2	299.4
Methionine	—	161.6	147.6	169.2
Phenylalanine	72.2	87.1	71.6	81.8
Proline	—	—	362.4	448.1
Threonine	67.3	77.8	70.1	93.0
Tyrosine	—	77.8	39.5	44.6
Valine	—	98.9	87.7	98.5
Total			2116.6	[2274.0

observed at a higher temperature. Table XIII shows this effect as observed in *Deliphila euphorbiae*.

The end of diapause, due to the circulation of ecdysone released from the prothoracic gland under the influence of the brain hormone, is accompanied by increased respiration and a rise in respiratory pigment turnover and concentration. As shown in Table XV, it is also accompanied by a progressive change of pattern of the aminoacidemia, ranging from a pattern characterized by a high proportion of lysine, arginine, alanine, proline, histidine and total

TABLE XV

Sphinx ligustri, NYMPHS; CONCENTRATION OF APPARENT AMINO ACIDS (mg/100 ml OF HEMOLYMPH PLASMA)

(Bricteux-Grégoire et al., 1960)

	Start	In cold room (5° C)					In cold glass-house (14° C)			
	9-11-55	20-11-55	20-12-55	9-2-56	27-3-56	17-5-56	20-11-55	20-12-55	9-2-56	27-3-56
Alanine	69.7	114.1	264.5	276.0	170.2	231.3	137.1	198.9	232.1	184.0
Arginine	286.0	239.6	230.3	275.0	148.8	288.6	242.3	253.1	257.9	308.8
Aspartic acid	4.0	7.1	7.1	11.2	17.9	9.9	10.2	5.1	10.6	12.9
Glutamic acid	299.7	149.1	58.0	125.8	138.6	259.1	170.7	205.1	129.9	223.9
Glycine	74.5	54.1	45.2	76.9	64.6	58.8	56.1	47.6	63.2	43.6
Histidine	142.0	118.8	129.1	126.3	63.8	103.5	130.1	120.9	98.4	147.2
Isoleucine	60.9	48.1	46.1	59.5	50.2	56.2	61.0	70.8	72.1	62.6
Leucine	73.7	61.2	55.6	71.8	64.6	61.5	77.6	85.6	84.5	84.0
Lysine	350.0	318.8	297.9	334.1	229.6	334.9	325.2	341.8	303.2	386.5
Methionine	79.3	87.9	81.4	77.4	61.6	99.1	97.6	98.2	72.8	85.9
Phenylalanine	27.2	36.2	36.5	32.1	32.3	35.0	32.2	39.1	39.0	35.0
Proline	196.0	164.4	170.8	154.8	105.9	152.0	174.0	200.2	196.9	192.8
Threonine	80.1	67.7	68.3	70.3	61.2	64.1	78.0	73.7	78.6	79.1
Tyrosine	7.2	22.0	18.3	14.8	7.7	10.6	10.2	7.4	10.6	32.5
Valine	100.2	90.3	90.5	103.9	86.7	93.2	102.4	113.3	104.0	108.0
Total	1850.5	1579.4	1599.6	1809.9	1348.7	1857.8	1704.7	1860.8	1753.8	1986.8

glutamic acid, to a pattern with two peaks: those of total glutamic acid and tyrosine. The change in the steady state of each amino acid is due to new conditions brought about by the circulation of ecdysone, and the liberation and incorporation of the amino acids concerned. In the case of glycine, the turnover of the free amino acids from the hemolymph to the adult tissues is higher in nymphs of *Sphinx ligustri* during their progress toward adulthood than in nymphs still in diapause (Bricteux-Grégoire *et al.*, 1960).

The last segment of the protostomian branch of animal phylogeny, formed by the class of insects, shows a number of taxonomic characteristics not present in their immediate ancestors. Some of these characteristics, linked to aspects of carbohydrate metabolism, have been discussed in this chapter and are defined as extensions to the pathways of carbohydrate metabolism.

BIBLIOGRAPHY

ANDREWARTHA, H. G., Diapause in relation to the ecology of insects, *Biol. Rev. Cambridge Phil. Soc.*, 27 (1952) 50–107.

BABERS, F. H., An analysis of the blood of the sixth-instar southern armyworm (*Prodenia eridania*), *J. Agr. Res.*, 57 (1938) 697–706.

BRICTEUX, S., T. FUKUDA, A. DEWANDRE AND M. FLORKIN, Contributions to silkworm biochemistry. VIII. Conversion of pyruvate into alanine, glycine and serine of silkfibroin, *Arch. Intern. Physiol. Biochim.*, 67 (1959) 545–552.

BRICTEUX-GRÉGOIRE, S., A. DEWANDRE, M. FLORKIN AND W. G. VERLY, Contributions à la biochimie du ver à soie. XI. Utilisation du carbone du formiate pour la biosynthèse des acides aminés de la fibroïne de la soie, *Arch. Intern. Physiol. Biochim.*, 67 (1959) 687–692.

BRICTEUX-GRÉGOIRE, S., G. DUCHÂTEAU-BOSSON, M. FLORKIN AND W. G. VERLY, Aspects biochimiques de la diapause nymphale des lépidoptères, *Arch. Intern. Physiol. Biochim.*, 68 (1960) 424–436.

BRICTEUX-GRÉGOIRE, S., CH. JEUNIAUX AND M. FLORKIN, Contributions à la biochimie du ver à soie. XXVIII. Biosynthèse de tréhalose à partir du pyruvate, *Arch. Intern. Physiol. Biochim.*, 72 (1964) 482–488.

BRICTEUX-GRÉGOIRE, S., CH. JEUNIAUX AND M. FLORKIN, Contributions à la biochimie du ver à soie. XXX. Biosynthèse de tréhalose et de glycogène à partir de glucose-1-phosphate. *Comp. Biochem. Physiol.*, 16 (1965) 333–340.

BÜCHER, TH., AND M. KLINGENBERG, The paths of hydrogen in the living organism, *Angew. Chem.*, 70 (1958) 552–570.

CANDY, D. J., AND B. A. KILBY, The biosynthesis of trehalose in the locust fat body, *Biochem. J.*, 78 (1961) 531–536.

CLEGG, J. S., AND D. R. EVANS, The physiology of blood trehalose and its function during flight in the blowfly, *J. Exptl. Biol.*, 38 (1961) 771–792.

DUCHÂTEAU, G., AND M. FLORKIN, Constitution de la composante protidique non-protéique de l'hémolymphe des chenilles et des chrysalides de *Sphinx ligustri* L., *Bull. Soc. Chim. Biol.*, 37 (1955) 239–245.

DUCHÂTEAU, G., AND M. FLORKIN, A survey of aminoacidemias with special reference to the high concentration of free amino acids in insect hemolymph, *Arch. Intern. Physiol. Biochim.*, 66 (1958) 573–591.

DUCHÂTEAU, G., AND M. FLORKIN, Sur la tréhalosémie des insectes et sa signification, *Arch. Intern. Physiol. Biochim.*, 67 (1959) 306–314.

DUCHÂTEAU-BOSSON, G., CH. JEUNIAUX AND M. FLORKIN, Contributions à la biochimie du ver à soie. XXVII. Tréhalose, tréhalase et mue, *Arch. Intern. Physiol. Biochim.*, 73 (1963) 566–576.

EVANS, D. R., AND V. G. DETHIER, The regulation of taste thresholds for sugars in the blowfly, *J. Insect Physiol.*, 1 (1957) 3–17.

FLORKIN, M., Taux des substances réductrices fermentescibles (glycémie vraie) du milieu intérieur des invertébrés, *Bull. Soc. Chim. Biol.*, 19 (1937a) 990–999.

FLORKIN, M., Variations de composition du plasma sanguin au cours de la métamorphose du ver à soie, *Arch. Intern. Physiol.*, 45 (1937b) 17–31.

FLORKIN, M., AND CH. JEUNIAUX, Hemolymph composition, in M. ROCKSTEIN (Ed.), *Physiology of Insecta*, Vol. 3, Academic Press, New York, 1964, pp. 109–152.

HELLER, J., AND A. MOKLOWSKA, Über die Zusammensetzung des Räupenblutes bei *Deilephila euphorbiae* und deren Veränderungen im Verlauf der Metamorphose, *Biochem. Z.*, 219 (1930) 473–489.

HEMMINGSEN, A. M., Blood sugar in the crayfish, *Skand. Arch. Physiol.*, 46 (1924) 204.

HOWDEN, G. F., AND B. A. KILBY, Trehalose and trehalase in the locust, *Chem. Ind. (London)*, (1956) 1453–1454.

KALF, G. F., AND S. V. RIEDER, The purification and properties of trehalase, *J. Biol. Chem.*, 230 (1958) 691–698.

LEVENBOOK, L., Fructose and the reducing value of insects' blood, *Nature*, 160 (1947) 465.

LEVENBOOK, L., The composition of horse botfly (*Gastrophilus intestinalis*) larva blood, *Biochem. J.*, 47 (1950) 336–346.

LUDWIG, D., Composition of the blood of Japanese beetle (*Popillia japonica* Newman) larvae, *Physiol. Zool.*, 24 (1951) 329–334.

MURPHY, T. A., AND G. R. WYATT, Enzymatic regulation of trehalose and glycogen synthesis in the fat body of an insect, *Nature*, 202 (1964) 1112–1113.

SAITO, S., Trehalose in the body fluid of the silkworm *Bombyx mori* L., *J. Insect Physiol.*, 9 (1963) 509–519.

SALT, R. W., Role of glycerol in the cold-hardening of *Bracon sephi* (Gahan), *Can. J. Zool.*, 37 (1959) 59–69.

SCHNEIDERMAN, H. A., AND C. M. WILLIAMS, The physiology of insect diapause VII. The respiratory metabolism of the cecropia silkworm during diapause and development, *Biol. Bull.*, 105 (1953) 320–324.

STEELE, J. E., The site of action of insect hyperglycemic hormone, *Gen. Comp. Endocrinol.*, 3 (1963) 46–52.

TODD, M. E., Concentration of certain organic compounds in the blood of the American cockroach, *Periplaneta americana* L., *J. N. Y. Entomol. Soc.*, 65 (1957) 85–88.

TODD, M. E., Blood composition of the cockroach, *Leucophaea maderae* Fabricius, *J. N. Y. Entomol. Soc.*, 66 (1958) 135–143.

TREHERNE, J. E., The absorption of glucose from the alimentary canal of the locust *Schistocerca gregaria* (Forsk.), *J. Exptl. Biol.*, 35 (1958a) 297–306.

TREHERNE, J. E., The absorption and metabolism of some sugars in the locust, *Schistocerca gregaria* (Forsk.), *J. Exptl. Biol.*, 35 (1958b) 611–625.

VON CZARNOWSKI, C., Zur papierchromatographischen Blutzuckerbestimmung bei der Honigbiene, *Naturwissenschaften*, 41 (1954) 577.

WILHELM, R. C., A study of the end products of anaerobic metabolism in pupae of the giant silkworm *Hyalophora cecropia* and *Samia cynthia*, Ph. D. Thesis, Cornell University, Ithaca, N.Y., 1960.

WILHELM, R. C., H. A. SCHNEIDERMAN AND L. J. DANIEL, The effects of anaerobiosis on the giant silkworms *Hyalophora cecropia* and *Samia cynthia* with special reference to the accumulation of glycerol and lactic acid, *J. Insect Physiol.*, 7 (1961) 273–288.

WILLIAMS, C. M., Morphogenesis and the metamorphosis of insects, *Harvey Lectures, Ser.* 47 (1951–1952) 126–155.

WYATT, G. R., AND G. F. KALF, Trehalose in insects, *Federation Proc.*, 15 (1956) 388.

WYATT, G. R., AND G. F. KALF, The chemistry of insect hemolymph. II. Trehalose and other carbohydrates, *J. Gen. Physiol.*, 40 (1957) 833–847.
WYATT, G. R., T. C. LOUGHHEED AND S. S. WYATT, The chemistry of insect hemolymph. Organic components of the hemolymph of the silkworm, *Bombyx mori*, and two other species, *J. Gen. Physiol.*, 39 (1956) 853–868.
WYATT, G. R., AND W. L. MEYER, The chemistry of insect hemolymph. III. Glycerol, *J. Gen. Physiol.*, 42 (1959) 1005–1111.
ZEBE, E. C., AND W. H. McSHAN, Trehalase in the thoracic muscles of the woodroach, *Leucophaea maderae*, *J. Cellular Comp. Physiol.*, 53 (1959) 21–29.

Chapter 9

Paleoproteins

Dead organisms are more or less rapidly putrefied according to conditions, and a large proportion of their substance is replaced by mineral deposits. Certain constituents nevertheless persist and are found in sediments as well as in sedimentary rocks.

Organic molecules have been detected in a number of fossils and their study is the subject matter of a new science: paleobiochemistry, or the biochemistry of fossils. Porphyrins have been isolated from coprolites of Eocene crocodiles by Fikentscher (1933), Abderhalden and Heyns (1933) have identified chitin in the remains of coleopterous wing-sheaths of the Eocene, and Abelson (1955, 1956, 1957a and b) has determined the existence of free amino acids in a series of fossils. According to Abelson, fossil shells older than 100 000 years do not contain protein remains*, but a mixture of free amino acids and short peptide chains, and after 25 million years only free amino acids persist. This opinion threatened to put an end to any hope of study of molecular phylogeny at the level of fossil remains, but fortunately more extensive studies have yielded preserved macromolecules of much greater age.

As can still be observed in the Onychophora such as *Peripatus*, the exoskeletons of primitive arthropods were essentially turgid tubes, the tensile forces of which were adapted to internal pressure, while, in a modern crustacean for example, mobility has been increased by reducing the internal turgidity and by developing the articulated exoskeleton which does not rely on internal pressure but on rigidity. In the primitive turgid tube—a complex of protein and chitin, a polymer of acetylglucosamine, resembling cellulose in many properties, was fitted for physiological adaptation, as it

---

* In a Pleistocene fossil shell of *Mercenaria mercenaria* about 100 000 years old, Jones and Vallentyne (1960) have detected a conchiolin liberating free amino acids by hydrolysis.

is at the same time strong and elastic, inextensible but flexible. Chitin is very insoluble, a property convenient for an exoskeleton, and its synthesis system is very common in monocellular organisms. It was therefore used, in combination with protein, to form the exoskeleton of primitive arthropods. Later on, rigidity was conferred upon the articulated skeleton either by sclerotization of the protein part of the protein–chitin exoskeleton or by impregnation with calcium carbonate. The first method consists in stabilizing the protein–chitin complex by the formation of bridges between free amino groups of the protein component (arthropodin) by allowing them to react with an orthoquinone. This method is compatible with flying, as it confers a rigid exoskeleton without increasing the weight too much. The second method has been used by the crustaceans of the subclass Malacostraca, in which parts of the protein–chitin exoskeleton are embedded in calcium carbonate. Stabilization by impregnation has also been used in the evolution of the exoskeleton of plants, primarily turgid tubes of cellulose, stabilized later on in the skeleton of trees by impregnation with lignin—a polymer of phenylpropane. In the exoskeletal material of vertebrates, keratin, the exoskeleton is made, not of cellulose or of a protein–chitin complex, but of two proteins, one of which, the fibrous α-keratin, derived from the tonofibrils of the epidermal cells, is of low sulfur content, and the other, amorphous and rich in sulfur, is the γ-keratin cement (see Matoltsy, 1962). Therefore cellulose, protein–chitin complex and α-keratin appear to be analogues providing the primary background of exoskeletons, the impregnation of which is secondarily insured by another series of analogous substances: lignine, orthoquinone, calcium carbonate and γ-keratin.

The exoskeleton of mollusks is typically represented by the shell, the original part of which is laid down by the mantle of the veliger larva. The extension takes place, in the adult, by a secretion at the edge of the shell. The prismatic layer is secreted by it under the horny periostracum. The successive layers accounting for the progressive thickening of the shell are produced by the whole surface of the mantle in the form of nacre. Several kinds of shell architectures are found in the mollusks: transverse shell plates in chitons; a single

shell, often coiled, in Gastropoda; a pair of hinged valves in pelecypods; a tubular shell in Scaphopoda; a spirally coiled shell formed
by a series of chambers (in the terminal of which the *Nautilus* lies);
a loose internal spiral in *Spirula*; a cuttle bone used as a density
regulator in *Sepia*. But whether external or internal, whether used
as a protecting armor, as a moisture-retaining device, as a swimming
apparatus, or in any other adaptation, the typical molluscan shell
is generally made up of three layers: cuticle, prisms and nacre.
The cuttle bone of *Sepia* still shows oblique calcareous particles
similar to the septa of the chamber skeleton of *Nautilus*, but it has
acquired a secondary structure in relation to its function as density
regulator.

Mollusks do not depart from the general method resorted to by
organisms in forming an exoskeleton. Theirs is composed of an
organic material impregnated with calcium carbonate. The organic
residue of the decalcification of a molluscan shell has been called
conchiolin by Frémy (1855).

In the layers of mother of pearl, the organic material is made up
of fibers of nacroin associated with a protein. Nacroin is essentially
composed of alanine and glycine, the nitrogen of these two amino

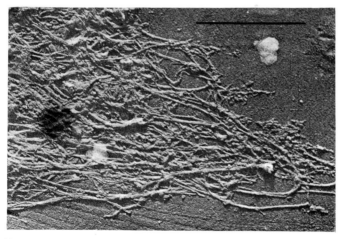

Fig. 41. *Pinctada margaritifera*. Nacroin fibers. Shadow cast with palladium;
× 28 000. Calibration: 1 μ. (Grégoire, Duchâteau and Florkin, 1955)

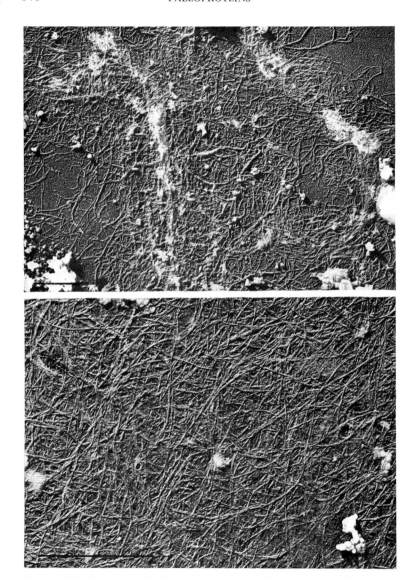

Fig. 46. *Vulsella vulsella.* Fibrillar residue of decalcification of a calcitic prism;
× 13 500. Calibration: 1 μ. (Grégoire, 1961a)
Fig. 47. *Nautilus pompilius* Lamarck. Residues of decalcification of the porcelain
layer of the shell wall (living chamber) consisting of fibrils; × 31 000. Calibration: 1 μ. (Grégoire, 1962)

Prism conchiolin surrounds each prismatic mineral structure. The prism conchiolin is, as in the case of nacreous conchiolin, built on a fibrillar web (Fig. 46), but the protein covering the latter has a very dense structure without lace-like appearance, except in *Mytilus*. While it is true that the lace-like structure of the nacreous conchiolin is always embedded in a mineral structure of aragonite crystals, aragonite itself is sometimes found in prism surrounded by the protein structure of prism. As is the case for some gastropods and *Nautilus*, it appears that, in the external layer of the shell, the prismatic layer is replaced by a very dense mineral structure—the porcelain component. This is deposited on a system of fibrils as shown in Fig. 47.

Organic matter can be maintained unmodified for very long periods of time, if the conditions are favorable. Abelson (1957a and b) has isolated free amino acids from a number of fossils, some as old as 450 million years, and he has discussed the conditions of maintenance of these molecules derived from the changes which took place with time in the proteins of the fossil. Amino acids such as alanine, aspartic acid, glutamic acid, glycine, leucine, proline and valine can endure more than a billion years at 25° C in anaerobic conditions. Fossil porphyrins have been isolated from sedimentary rocks (Treibs, 1934–1936). But the most striking recent discovery in the field is certainly the identification by Grégoire (1958b, 1959) of lace-like structures in fossil shells, similar to those described above in the case of nacreous conchiolin. When we were able to obtain enough material of such nacreous remains, we undertook a chemical study of them and found that the lace-like structures observed in fossils through the study of replicas and of decalcified material were actually conchiolin remains still presenting the amino acid pattern of nacreous conchiolin. But even more striking is the fact that the typical chemical composition of the conchiolin of mother-of-pearl has been found in fossil shells from the Eocene period (60 million years old).

The chromatograms shown in Fig. 48 relate to analogous weights of nitrogen remaining in fragments of mother-of-pearl (the ultra-structure of which is shown in Fig. 49) after washing free amino

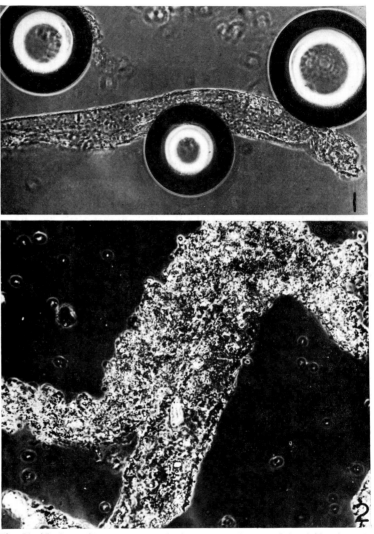

Fig. 51. (1) *Atrina* (*Pinna nigra* Durh. (recent). End stage of decalcification of a single prism. In the sheath shown here, shrinkage and wrinkling conceal the transverse striation. Phase contrast × 230. (Grégoire, 1961a). (2) *Pinna affinis* (London Clay, Lower Eocene, 60 million years). Elongated, folded and wrinkled fragments of tubular prism sheaths. The transverse striation is concealed in part by small mineral fragments resisting demineralization by the chelating agent used, and as in the recent sample, by shrinking of the fragments. Phase contrast × 310. (Bricteux-Grégoire, Florkin and Grégoire, unpublished)

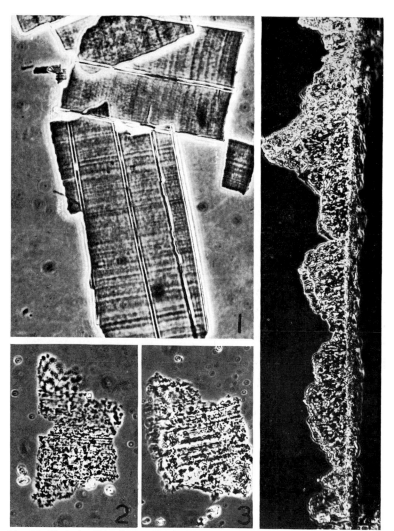

Fig. 52. (*1*) *Atrina* (*Pinna*) *nigra* Durh. (recent). Sheath fragments from decalcified prisms showing a transverse striation on rectangular facets. Phase contrast × 620. (Grégoire, 1961a). (*2*) and (*3*) *Pinna affinis* (London Clay, Lower Eocene, 60 million years). Debris of prism sheath facets showing small mineral granules disposed along the transverse striation (see *1*). Phase contrast × 310. (Bricteux-Grégoire, Florkin and Grégoire, unpublished). (*4*) *Inoceramus* sp. (Gault, Cretaceous, 135 million years). Elongated sheath shred left by decalcification of a prism. As in Fig. 51, *2*, and in this figure *2* and *3*, abundant microcrystals, escaping demineralization, persist, either attached to or embedded in the substance of the sheath. Phase contrast × 310. (Bricteux-Grégoire, Florkin and Grégoire, unpublished)

TABLE XVIII

COMPARISON OF MODERN PRISMS AND FOSSILS

(Bricteux-Grégoire, Florkin and Grégoire, unpublished)

| | Modern | | | | | | Fossil | | | | | |
| | Atrina nigra | | | Pinna nobilis | | | Pinna affinis | | | Inoceramus | | |
	μg/g	% N	mol. fr.	μg/g	% N	mol. fr.	μg/g	% N	mol. fr.	μg/g	% N	mol. fr.
Lysine	145	0.7	0.8	110	0.5	0.6	tr			1.58	0.7	4.8
Histidine	70	0.3	0.4	89	0.4	0.5	tr			1.13	0.5	3.2
Arginine	252	1.1	1.1	36	1.5	1.7	tr			0.96	0.4	2.4
Aspartic acid	1750	9.6	10.4	3920	20.5	23.3	23.4	2.6	11.1	3.03	1.6	10.0
Threonine	173	1.1	1.1	231	1.4	1.5	8.8	1.1	4.7	1.18	0.7	4.4
Serine	530	3.7	4.0	617	4.1	4.6	27.5	3.8	16.5	3.88	2.5	16.4
Glutamic acid	330	1.6	1.8	348	1.7	1.9	31.8	3.1	13.6	5.13	2.4	15.5
Proline	314	2.0	2.2	316	1.9	2.2	tr			tr		
Glycine	4390	42.4	46.2	3500	32.4	36.9	26.8	5.2	22.5	3.60	3.3	21.3
Alanine	530	4.3	4.7	640	5.0	5.7	20.0	3.3	14.1	1.62	1.2	8.1
Valine	930	5.8	6.3	833	4.9	5.6	9.2	1.1	5.0	1.30	0.8	4.9
Isoleucine	334	1.8	2.0	558	3.0	3.4	7.9	0.9	3.8	0.99	0.5	3.4
Leucine	1220	6.8	7.4	740	3.9	4.5	12.2	1.4	5.9	1.69	0.9	5.7
Tyrosine	2130	8.6	9.3	1200	4.6	5.3	tr			tr		
Phenylalanine	472	2.1	2.3	508	2.1	2.4	7.6	0.7	2.9	tr		
Total		91.9	100.0		87.9	100.1		33.2	100.1		72.2	100.1
Amino nitrogen	1930			2020			96			20		
Ammonia	70	3.0		90	3.7		131	113		14.1	56.7	

water in order to remove soluble components, such as free amino acids. The insoluble fossil remnants were then dried, weighed and hydrolyzed by 6 N HCl for 24 h under reflux. After removing HCl by evaporation, the different amino acids contained in the hydrolysates were identified by column chromatography on a Beckman Spin-

TABLE XIX

AMINO ACIDS IN HYDROLYSATES OF GRAPTOLITES, AFTER DECALCIFICATION AND WASHINGS[a]

(Foucart et al., 1965)

	Pristiograptus gotlandicus and P. dubius (Silurian)		Monograptidae gn. sp. (Silurian)		Climacograptus typicalis (Ordovician)	
	$\mu g/g$	% mol. fr.	$\mu g/g$	% mol. fr.	$\mu g/g$	% mol. fr.
Aspartic acid	218	9	560	8.6	68	10
Threonine	108	4.9	290	4.9	26	4.3
Serine	214	11	550	10.6	122	22.8
Glutamic acid	380	13.9	1100	15.3	96	12.8
Proline	(83)	(3.9)	(340)	(6)	tr	—
Glycine	280	20.1	760	20.8	89	23.4
Alanine	103	6.3	410	9.5	40	8.9
Valine	116	5.3	(240)	(4.1)	tr	—
Isoleucine	97	4	230	3.6	(15)	2.3
Leucine	190	7.8	390	6.1	(13)	(2)
Tyrosine	(42)	(1.2)	(90)	(1.1)	—	—
Phenylalanine	(87)	(2.9)	(110)	(1.3)	—	—
Lysine	96	3.5	(310)	(4.4)	70	9.4
Histidine	60	2.2	(84)	(1.1)	29	3.7
Arginine	128	4	(140)	(2.2)	tr	—
Ammonia	223					

[a] The amounts indicated in brackets could only be calculated approximately on account of their low values.

Bibliography p. 155

co automatic apparatus. The results obtained are shown in Table XIX.

It can be seen that, in addition to high proportions of ammonia, amino acids are present in the hydrolysates of the three samples studied (*Pristiograptus*: 2,202 μg/g; Monograptidae: 5,604 μg/g; *Climacograptus*: 568 μg/g).

As the soluble material of the graptolites was removed by repeated treatments, we can consider that the amino acids contained in the hydrolysates were of proteic origin. Identification of peptide linkages by means of the biuret reaction, as used by Florkin *et al.* (1961) in the case of conchiolin remnants, was not possible on account of the dark color of the graptolite fragments.

The question rises as to whether these proteins were contained in the graptolites themselves or originated in the sediment from which the graptolites were extracted. The analysis of one graptolitic rock as compared with the analysis of the graptolites extracted from this rock made it possible to eliminate this hypothesis. The results given in Table XX show that the amino acid concentrations in the rock were much lower than those in the graptolites. On the other hand, some amino acids present in the graptolites were not detected in the sediments. The organic composition of the graptolite remnants is thus quantitatively and qualitatively distinct from the kerogen of the sedimentary rock.

It was also possible to rule out the possibility of contamination of the graptolites themselves. Indeed, examination of the fragments under the electron microscope did not reveal any appreciable level of typical contamination by bacteria, algal filaments, conchiolin remnants and so on. On the other hand, parallelism between the results obtained for the three samples of different origin and geological age could hardly be consistent with the possibility of contamination by exogenous organic material.

In conclusion, the amino acids observed in the hydrolysates of washed and decalcified fragments of graptolites can reasonably be considered to have originated in residual fossil proteins derived from the living animals. In all three different species of graptolites so far examined, these proteins show a similar amino acid pattern, characterized by high amounts of serine (molecular fraction:

TABLE XX

COMPARISON BETWEEN THE AMOUNTS OF AMINO ACIDS OF PROTEIN ORIGIN IN
GRAPTOLITES EXTRACTED FROM A ROCK WITH THOSE FOUND IN THE
ROCK ITSELF

(Foucart et al., 1965)

	$\mu g/g$		Molecular fraction	
	Graptolites[a]	Rock[b]	Graptolites	Rock
Aspartic acid	68	12.8	10	7.4
Threonine	26	8	4.3	(5.1)
Serine	122	45.4	22.8	33.4
Glutamic acid	96	9.6	12.8	5.1
Proline	tr	—	tr	—
Glycine	89	30	23.4	30.7
Alanine	40	12	8.9	10.4
Valine	tr	—	tr	—
Isoleucine	(15)	5.6	2.3	(3.3)
Leucine	(13)	7.4	(2)	(4.3)
Tyrosine	—	—	—	—
Phenylalanine	—	—	—	—
Lysine	70	—	9.4	—
Histidine	29	—	3.7	—
Arginine	tr	—	tr	—
Total	568	130.8	99.6	99.7

[a] Climacograptus typicalis.
[b] Ordovician limestone containing Climacograptus typicalis (from Ohio) after
decalcification and removal of the Graptolite fragments.

10.6–22.8), alanine (6.3–9.5), glycine (20.1–23.4), aspartic acid
(8.6–10) and glutamic acid (12.8–15.3). Such high total amounts
of glycine, serine and alanine suggest that these graptolitic proteins
are of a scleroprotein nature. The study of this protein material with
the electron microscope is being continued by Dr. Ch. Grégoire
in our laboratory.

Bibliography p. 155

Finally, in contrast to the beliefs of some investigators (*e.g.* Kraft, 1926; Manskaia and Drozdova, 1962), the three samples examined did not contain any trace of chitin. This polysaccharide has not been detected by the specific enzymatic method of Jeuniaux (1963), and the chromatograms of the hydrolysates did not reveal the presence of glucosamine. Furthermore, tests on the graptolites have not revealed any trace of cellulose.

All the data on paleoproteins reviewed in this chapter show that fossils may contain remains of proteins which are not the result of contamination either during the life of the organism concerned or after its death. This conclusion strengthens the hope that we shall be able to isolate from such paleoproteins definite units which we shall be able eventually to compare, with respect to their primary structure, with corresponding proteins in their distant progeny.

BIBLIOGRAPHY

ABDERHALDEN, E., AND K. HEYNS, Nachweis von Chitin in Flügelresten von Coleopteren des oberen Mitteleocäns (Fundstelle Geiseltal), *Biochem. Z.*, 259 (1933) 320–321.

ABELSON, P. H., Paleobiochemistry. Organic constituents of fossils, *Carnegie Inst. Wash. Yearb.*, 54 (1955) 107–109.

ABELSON, P. H., Paleobiochemistry, *Sci. Am.*, 195, No. 1 (1956) 83–92.

ABELSON, P. H., Some aspects of paleobiochemistry, *Ann. N.Y. Acad. Sci.*, 69 (1957a) 276–285.

ABELSON, P. H., Organic constituents of fossils, *Geol. Soc. Am.*, *Mem.* 67 (1957b) 87–92.

FIKENTSCHER, R., Koproporphyrin in tertiären Krokodilkot, *Zool. Anz.*, 103 (1933) 289–295.

FLORKIN, M., CH. GRÉGOIRE, S. BRICTEUX-GRÉGOIRE AND E. SCHOFFENIELS, Conchiolines de nacres fossiles, *Compt. Rend.*, 252 (1961) 440–442.

FOUCART, M. F., S. BRICTEUX-GRÉGOIRE, CH. JEUNIAUX AND M. FLORKIN, Fossil proteins of graptolites, *Life Sciences*, 4 (1965) 467–471.

FRÉMY, E., Recherches chimiques sur les os, *Ann. Chim. et Phys.*, (*Paris*), [3] 43 (1855) 47–107.

GRANDJEAN, J., CH. GRÉGOIRE AND A. LUTTS, On the mineral components and the remnants of organic structures in shells of fossil molluscs, *Bull. Classe Sci. Acad. Roy. Belg.*, [5] 50 (1964) 562–595.

GRÉGOIRE, CH., Topography of the organic components in mother-of-pearl, *J. Biophys. Biochem. Cytol.*, 3 (1957) 797–808.

GRÉGOIRE, CH., Sur la structure, étudiée au microscope électronique, des constituants organiques du calcitostracum, *Arch. Intern. Physiol. Biochim.*, 66 (1958a) 658–661.

GRÉGOIRE, CH., Essai de détection au microscope électronique des dentelles organiques dans les nacres fossiles (ammonites, céphalopodes, gastéropodes et pélécypodes), *Arch. Intern. Physiol. Biochim.*, 66 (1958b) 674–676.

GRÉGOIRE, CH., A study on the remains of organic components in fossil mother-of-pearl, *Bull. Inst. Roy. Sci. Nat. Belg.*, 35, No. 13 (1959) 1–14.

GRÉGOIRE, CH., Further studies on structure of the organic components in mother-of-pearl, especially in Pelecypods (Part I), *Bull. Inst. Roy. Sci. Nat. Belg.*, 36 No. 23 (1960a) 1–22.

GRÉGOIRE, CH., Sur la structure submicroscopique de la conchioline associée aux prismes des coquilles de mollusques. *Bull. Inst. Roy. Sci. Nat. Belg.*, 37, No. 3 (1961a) 1–34.

GRÉGOIRE, CH., Structure of the conchiolin cases of the prisms in *Mytilus edulis* L., *J. Biophys. Biochem. Cytol.*, 9 (1961b) 395–400.

GRÉGOIRE, CH., On submicroscopic structure of the *Nautilus* shell, *Bull. Inst. Roy. Sci. Nat. Belg.*, 38, No. 49 (1962) 1–71.

GRÉGOIRE, CH., G. DUCHÂTEAU AND M. FLORKIN, La trame protidique des nacres et des perles, *Ann. Inst. Océanog. (Monaco)*, 31 (1955) 1–36.

JEUNIAUX, CH., *Chitine et Chitinolyse. Un Chapitre de la Biologie Moléculaire*, Paris, Masson, 1963.

Jones, J. D., and J. R. Vallentyne, Biogeochemistry of organic matter. I. Polypeptides and amino acids in fossils and sediments in relation to geothermometry, *Geochim. Cosmochim. Acta*, 21 (1960) 1–34.

Kraft, P., Ontogenetische Entwicklung und Biologie von Diplograptus und Monograptus, *Palaeont. Z.*, 7 (1926) 207–249.

Manskaia, S. M., and T. V. Drozdova, Transformation of organic compounds in sedimentary rocks and the organic matter of graptolites of dictionemae shales, *Geokhimiya*, (1962) 952–962.

Matoltsy, G., Structural and chemical properties of keratin forming tissues, in M. Florkin and H. S. Mason (Eds.), *Comparative Biochemistry*, Vol. IV, Academic Press, New York, 1962, pp. 343–369.

Roche, J., G. Ranson and M. Eysseric-Lafon, Sur la composition des scléroprotéines des coquilles des mollusques (conchiolines), *Compt. Rend. Soc. Biol.*, 145 (1951) 1474–1477.

Treibs, A., Über das Vorkommen von Chlorophyllderivaten in einem Ölschiefer aus der oberen Trias, *Ann. Chem.*, 509 (1934a) 103–114.

Treibs, A., Chlorophyll- und Häminderivate in bituminösen Gesteinen, Erdölen, Erdwachsen und Asphalten. Ein Beitrag zur Entstehung des Erdöls, *Ann. Chem.*, 510 (1934b) 42–62.

Treibs, A., Chlorophyll- und Häminderivate in bituminösen Gesteinen, Erdölen, Kohlen, Phosphoriten, *Ann. Chem.*, 517 (1935a) 172–196.

Treibs, A., Porphyrine in Kohlen, *Ann. Chem.*, 520 (1935b) 144–150.

Treibs, A., Chlorophyll- und Häminderivate in organischen Mineralstoffen, *Angew. Chem.*, 49 (1936) 682–686.

Trim, A. R., Studies in the chemistry of the insect cuticle. I. Some general observations on certain Arthropod cuticles with special reference to the characterization of the proteins, *Biochem. J.*, 35 (1941) 1088–1098.

Wetzel, G., Die organischen Substanzen der Schalen von *Mytilus* und *Pinna*, *Z. Physiol. Chem.*, 29 (1900) 386–410.

Chapter 10

Evolving Organisms and Molecules

As Simpson (1964, 1965) has pointed out, before the discovery of DNA, the present author (Florkin, 1944) presented and discussed the systematics and evolution of certain families of molecules. By a phylogeny of macromolecules we do not mean a phyletic evolution that could be followed for example at the level of a population, although we have some hopes of attacking phyletic evolution from the molecular point of view by pursuing a study of paleoproteins (Chapter 9). For the time being, though, we are only concerned with what is generally called divergent evolution.

When we study the phylogeny of macromolecules, we compare homologous macromolecules along the branches of the phyletic tree. Homology, as defined in this book (Chapter 2), starts from the existence of a common initial prototype at the level of base sequences in the nucleic acids of the genes and in the more or less direct replicas of these sequences in the form of amino acid sequences in proteins. The test of this definition is a high degree of isology in the primary structure of nucleic acids and proteins. Homology, as we have said, can be extended to identical or similar chains of homologous enzymes, as well as to the result of a biosynthetic process governed by such a chain of homologous enzymes. However, when the end-point is neither nucleic acid nor protein, isology is not a decisive test of homology, which can only be recognized when the chain of enzymes catalyzing the steps of the biosynthetic pathway is homologous. The homology lies at the level of the protein catalysts, and therefore the homology of the end-points is of an indirect nature, as opposed to the direct homologies obtaining in the cases of nucleic acids and proteins. The concepts of direct and indirect homology correspond respectively to the denominations of semantic molecules (semantides) and episemantic molecules (episemantides) as proposed by Zuckerkandl and Pauling (1964).

Bibliography p. 165

Are we competent to compare homologous (highly isologous) macromolecules of nucleic acids or proteins, or indirectly homologous molecules, along the branches of the phylogeny proposed by the biologists?

Several objections can be raised. When we compare a macromolecule in an organism living today with a homologous molecule in another organism, also living today, but belonging to a more primitive category according to the sequence in the phylogenic tree, do we not unduly ignore the fact that the more "primitive" species have of course also changed since the branch on which we find the more "specialized" species separated? A comparative study of the primary structures of macromolecules of hemoglobins, of cytochromes, and of insulins, as reported in Chapter 3, has nevertheless indicated that the more the organisms considered are far away on the branches of phylogeny, the more different are the primary structures of the macromolecules concerned. All organisms have changed in the course of time, but those which have changed less from the organismic viewpoint are also those which have changed less from the molecular viewpoint. To this, the case of sibling species can apparently be opposed. In these species, the genotype has varied, while it appears that the morphological characteristics have remained almost identical; these species are sympatric, but reproductively isolated. It can certainly be conceived that, in sibling species, a small number of biochemical characteristics may have changed while the morphology has not, but, in fact, a deeper study has revealed morphological differences among sibling species (see Mayr, 1963). As we have stated, the opportunity for more direct phyletic comparison will be afforded by paleobiochemical studies (Chapter 9).

On the other hand, it can be objected that the phylogeny drawn by naturalists is of an unreliable nature. Granted that it rests on a large collection of anatomical, embryological and paleontological data, nobody accepts it as expressing an absolute truth. In taking the tree of phylogeny as our guide, we choose the best available information. It has happened that biochemists, in their comparative studies, have noted gross discrepancies between organismic and molecular data. In certain cases at least, these discrepancies have been taken into account by competent biologists in the process of a revision of their phyletic sequences of organisms, and it has happened that this revision has brought a change in the tree of phylogeny.

In the study of phylogeny of proteins, enquiries have mainly been directed so far to the evolution of the carrier of a definite biochemical property. Cytochromes, hemoglobins, insulins, etc. have been compared in phylogeny. How much more profitable it would be to follow not a property and its carrier, but a primary sequence and eventually its changes in properties. This will be possible only when we have at our disposal a large number of data on primary sequences in the chains of a large number of proteins and of nucleic acids. It would also be of the most pregnant interest to be able to compare, in a short segment of a branch in phylogeny, the parallel changes, for instance of one of the chains of hemoglobin, and of the chains of purine and pyrimidine bases on the appropriate gene corresponding to the amino acid primary structure concerned.

With respect to the substances displaying indirect homology (episemantides of Zuckerkandl and Pauling), this kind of homology can only be recognized if we know that the biosynthetic pathways are the same and that the enzymes involved are homologous in the pathways compared. There are of course, as we have noted in Chapter 4, many examples of convergence in the case of isologous molecules other than nucleic acids and proteins. This means that isologous molecules of this category can eventually result from different biosynthetic pathways, not catalyzed by homologous enzyme chains and, consequently, can be non-homologous. It is true, as Wald (1963) writes, that if the probability is remote of evolution producing the same organ twice (and we may add: the same primary protein structure), the probability is greater for the convergent production of non-nucleic and non-protein molecular structures. This is certainly true, and we have called attention to it by distinguishing direct homology (nucleic acids, proteins) and indirect homology (other molecules). In his remarkable studies on retinal pigments, Wald described the nature and zoological distribution of retinenes. The type of retinene (see Chapter 4) is, in view of the

Bibliography p. 165

observations made by Bridges (1965), an accommodation resulting from the intensity of light acting upon the eye and of the exposure time. Bridges has been able to show that, in the peociliid fish *Belonesox belizanus*, the light quantity (intensity + exposure time) brings about changes in the proportions of retinene$_1$ and retinene$_2$, both present in the eyes of this species. It is therefore safer to avoid phyletic implications in the case of retinal pigments. On the other hand, the retinenes derive from the carotenoids of the food, by the intervention of enzyme systems about which we do not have comparative data so far. The retinal pigment appears to be of an ecophenotypical nature.

Convergences can be detected at the level of the phylogeny of a definite locus of a primary structure, as Zuckerkandl (1965) has described in a study on hemoglobin. Nevertheless, this does not prevent a definition of the behavior of the whole sequence in phylogeny.

Descent with change in structure can be detected at the level of primary structures of proteins, as shown in Chapter 3, and it may be hoped that data will also be gathered in the near future with respect to descent with change at the level of the primary structures of nucleic acids. Concerning the phylogeny of biosynthetic chains, we have seen, in Chapter 4, that the changes in descent can often be detected in the form of extensions on metabolic lines. The possibility that an enzyme added in this way is homologous with its immediate predecessor, in the way common to the hemoglobin chains developed in succession in phylogeny, can be suggested, but has not so far been experimentally tested.

When indirectly homologous molecules are considered, what is really homologous is each protein composing the biosynthetic chain. When one or several enzymes are, in a following phyletic stage, added to the same chain, the new product can be considered to descend, in the phylogeny of the organisms concerned, from the previous products. It is in this context that it may be proposed that, in a number of cases at least, the steps of biosynthesis of molecules repeat the phylogeny of the molecules, as defined above.

Whether or not the primary structure of a protein is modified in the phylogeny of the protein, the *properties* of this protein can be modified. With reference to ribonuclease (see Chapter 3), it has been customary to consider that the tertiary structure of the protein is controlled by its primary structure, but our more recent knowledge of trypsin and chymotrypsin points to their homology in spite of their different specificities in terms of enzymatic activity. This points to the possibility of differences in tertiary structure and enzymatic activity of very isologous proteins. On the other hand, our present knowledge of dehydrogenases (see Whitehead, 1965) shows that a change of quaternary structure can change the specific catalytic properties of a definite amino acid sequence. While therefore considering the changes of primary structure in the phylogeny of proteins, we can leave open the consideration of properties, including tertiary and quaternary structures as well as catalytic properties or the property of reversibly combining with oxygen, etc. It appears premature to go more specifically into the possible modifications of these *properties* in phylogeny; this problem remains a goal for future work.

It happens that the addition of an enzyme or of a few enzymes changes the nature of a molecular unit. An enzyme may also be lost at one or other step in the phylogeny of organisms. This is the case, as we have seen, with the enzyme chitin-synthetase, and the result of this has been that the biosynthesis of chitin is lost at the base of the branch of deuterostomians (Chapter 5).

In comparative biochemistry in its broader sense, data related to the concentration of this or that component of a body fluid or of a category of cells are often compared. These concentrations are, in fact, the translation of the existence of a steady state in each case. These aspects are more complicated than those of metabolic pathways, in the sense that the concentration of a definite component may represent a steady state at the crossroads of several metabolic pathways. It is nevertheless only our ignorance which prevents us from relating changes of chemical composition to the phylogeny of organisms.

In Chapter 7, the phylogeny of steady states expressed in concentrations of inorganic ions in the hemolymph of insects is considered.

Bibliography p. 165

Some aspects of the free-amino-acid concentrations are also discussed in Chapters 7 and 8 in their relation to insect phylogeny.

The changes of structure and the changes of properties (as defined above) that can be detected at the level of constituent molecules and macromolecules along the paths of the phylogeny of organisms containing these molecules and macromolecules are aspects of adaptive radiations.

In order to relate the consideration of the molecular or macromolecular level of organization with the level of the organisms themselves, it is important to consider, the molecular or macromolecular units whether they do or do not undergo, in the descent of organisms, changes in structure and/or properties, in the framework of the functions as considered by physiologists. They may undergo changes of function at a level higher than the molecular one—a change which can be called a *physiological radiation* (Florkin, 1957, 1959, 1960, 1962, 1963). Simpson (1949) defines adaptive radiation in its descriptive and organismic aspect, as the diversification of a group, in the course of its evolution, in all the directions which are permitted by its own possibilities and by the media with which it comes into contact.

By the expression *physiological radiation* we mean the diversification of a molecule, macromolecule or enzyme system, whether or not it undergoes changes in structure and/or properties, in all the physiological directions, *i.e.* in "functions" characteristic of levels higher than the molecular level*. A biochemical system may show a physiological radiation involving changes in structure and properties of the constituents of the system. Snakes, for example, do not mix digestive secretions with their prey in a process of mastication. They swallow their prey after having injected it with a secretion which initiates hydrolysis. In the least-specialized form, for example in *Colubridae opisthoglyphae*, a simple secretory tooth appears at the rear of the upper jaw and serves for the injection of a

---

* The concept of physiological radiation also involves a consideration of the phyletic relations of the organisms concerned and, in this respect, differs from the concept of *heterotypic expressions* proposed by Mason (1955).

secretion the function of which is purely digestive. In more specialized forms, this organ, following a decrease in length of the maxilla, approaches the anterior part of the buccal cavity and becomes an aggressive and defensive organ, as is the case in *Colubridae protero-glyphae* and even more so in the *Viperidae*. The digestive origin of the secretion is further borne out by the presence in snake venoms of such hydrolases as proteases, peptidases, phosphatases, esterases and lecithinases. The new specialization expresses itself by the presence of hyaluronidase, assuring the diffusion of the venom, and by the presence of substances of high toxicity (see Zeller, 1948).

   The pentose cycle which, in a definite context, exercises the function of metabolizing glucose and of biosynthetizing pentoses, is used in vertebrate red blood cells for the continuous reduction of methemoglobin.

   Bioluminescence may be regarded as a vestigial mechanism for the detoxication of oxygen, which appeared, during the evolution of primitive organisms, in an atmosphere progressively richer in oxygen (McElroy and Seliger, 1963). As the organisms became aerobic, luminescence lost its selective characteristic and generally disappeared. In a number of cases, though, it has undergone a physiological radiation which is to be seen, for example, in the role it plays in the identification of the female glow-worm.

   The system of chitinolytic enzymes is used, in the majority of invertebrates, in the digestion of their food chitin, derived from the exoskeleton of the prey, with the liberation of acetylglucosamine, which is absorbed in the digestive tract (see Chapter 5). On the other hand, in vertebrates, a single enzyme of the chitinolytic system is retained at the level of the digestive tract, chitinase, and the effect is that the chitin of preys is hydrolyzed without liberation of acetylglucosamine. The sole effect of this is to allow the penetration of the enzymes into the food particles. Other physiological radiations of the chitinolytic system result from the use of this system at the ectodermal level: these radiations are active in arthropod molting and in the eclosion of nematode eggs. The enzymes of the chitinolytic system play a role in certain cases of parasitism. *Beauveria bassiana*, a parasitic mycete, is equipped with the system of chitinolysis

Bibliography p. 165

through which it is able to penetrate the host cuticle and to develop inside its tissues.

If it is true that a biochemical system can show a physiological radiation, it may also happen that a biochemical constituent, without apparently presenting any chemical modification, be introduced into a new biochemical system (Florkin, 1957, 1959, 1960). The mechanisms of hormonal regulation provide us with many instances of the kind of physiological radiation consisting of the introduction of a biochemical constituent into a new biochemical system. The secretion of milk, due to the biochemical differentiation of one type of mammalian cell, is provoked and controlled by prolactin, resulting from the biochemical specialization of another type of cell, in the adenohypophysis of fish, amphibians and reptiles. Its intervention in the secretion of milk in mammals is thus an insertion into a new biochemical system. Another example of the same type is the action of pitocin on the mammalian uterus. The hormone is present in all vertebrates and acts in the control of water metabolism. Its action on mammalian uterus demonstrates its insertion into a more specialized system.

A more detailed knowledge of the nature of physiological radiations of molecules, macromolecules and enzyme systems will supply a host of information about the common biochemical background of apparently different physiological aspects and will therefore contribute to building comparative physiology, not as it is now considered, on analogies, but according to the evolutionary aspects of homologies.

From the enumeration of the aspects of molecular evolution described above, it may be concluded that the change of primary protein structure with descent is only one among many. Molecular evolution may take many other forms. But to systematize all these forms, the student of molecular evolution has no other possible methodology than to consider molecular changes of all kinds along the sequences of organisms in phylogeny.

In conclusion, it may be stated that the opposition sometimes established between the organismic and molecular approach to evolution is without meaning.

BIBLIOGRAPHY

BRIDGES, C. D. B., Visual pigments in a fish exposed to different light-environments, *Nature*, 206 (1965) 1161–1162.

FLORKIN, M., *L'Évolution Biochimique*, Masson, Paris, 1944.

FLORKIN, M., Biochimie et évolution animale, *Actes Soc. Helv. Sci. Nat.*, 136 (1957) 35–57.

FLORKIN, M., L'extension de la biosphère et l'évolution biochimique, in F. CLARK AND R. L. M. SYNGE (Eds.), *The Origin of Life on the Earth*, Pergamon, London, 1959, pp. 503–515.

FLORKIN, M., *Unity and Diversity in Biochemistry*, Pergamon, London, 1960.

FLORKIN, M., Isologie, homologie, analogie et convergence en biochimie comparée, *Bull. Classe Sci. Acad. Roy. Belg.*, [5] 48 (1962) 819–824.

FLORKIN, M., L'évolution biochimique et la radiation physiologique des systèmes biochimiques chez les animaux, in A. I. OPARIN (Ed.), *Evolutionary Biochemistry*, Pergamon, London, 1963, pp. 250–270.

MCELROY, W. D., AND H. H. SELIGER, Origin and evolution of bioluminescence, in A. I. OPARIN (Ed.), *Evolutionary Biochemistry*, Pergamon, London, 1963, pp. 158–168.

MASON, H. S., Comparative biochemistry of the phenolase complex, *Advan. Enzymol.*, 16 (1955) 105–184.

MAYR, E., *Animal Species and Evolution*, Harvard University Press, Cambridge, 1963.

SIMPSON, G. G., *The Meaning of Evolution*, Yale University Press, 1949.

SIMPSON, G. G., Organisms and molecules in evolution. Studies of evolution at the molecular level lead to greater understanding and a balancing of viewpoints, *Science*, 146 (1964) 1535–1538.

SIMPSON, G. G., Organisms and molecules in evolution, in H. PEETERS (Ed.), *Protides of the Biological Fluids*, Vol. 12, Elsevier, Amsterdam, 1965, pp. 29–35.

WALD, G., Phylogeny and ontogeny at the molecular level, in A. I. OPARIN (Ed.), *Evolutionary Biochemistry*, Pergamon, London, 1963, pp. 12–51.

WHITEHEAD, E. P., A theory of the quaternary structure of dehydrogenases, dehydrogenating complexes and other proteins, *J. Theoret. Biol.*, 8 (1965) 276–306.

ZELLER, E. A., Enzymes of snake venoms and their biological significance, *Advan. Enzymol.*, 8 (1948) 459–495.

ZUCKERKANDL, E., Further principles of chemical paleogenetics as applied to the evolution of hemoglobin, in H. PEETERS (Ed.), *Protides of the Biological Fluids*, Vol. 12, Elsevier, Amsterdam, 1965, pp. 102–109.

ZUCKERKANDL, E., AND L. PAULING, in *Problems of Evolutionary and Technical Biochemistry*, Dedicated to Acad. A. I. OPARIN, Science Press, Acad. Sci. U.S.S.R., (in Russian), p. 54 [cited by E. Zuckerkandl, 1965]. [The original English text was published in 1965: Molecules as documents of evolutionary history, *J. Theoret. Biol.*, 8 (1965) 357–366].

Subject Index

Acetylglucosamine, 51, 58, 60, 163
—, product of chitinase action, 51
N-Acetyl-D-glucosamine, diholoside of, 50
ACTH, amino acid sequence, 11
—, primary structure, 7
—, properties associated with specific antigenic reactions, 16
Adaptive radiations, 48, 162
Adenase, 74
Aldolase I, 48
Aldolase II, 48
Amino acid(s), biosynthesis of sequences, 5
—, catabolism, 83
— composition, of conchiolin from mother-of-pearl, 144
—, free, from fossils, 141
— metabolism, ammonia as end-product, 75
— pool of hemolymph, 112
— synthesis, in insects, 118, 119
Aminoacidemia, in insects, 89
—, modification of, in insects, 103
—, in nymphs and in caterpillars of Sphinx ligustri, 124
—, pattern, change of, 127, 128
Ammonia excretion, 65
Ammonia pool, in biosynthesis of purines, 72
Ammonia, toxicity of, 62
Ammonia toxicity, and excretory N products, 63
Ammoniogenesis α, 62
Ammoniotelic nitrogen metabolism, 81
Amphibia, chitinolytic system, 57
—, in fresh water, osmotic pressure and plasma composition, 69
—, osmoregulation, 67

Amphibia, (continuation)
—, osmotic pressure, 68
—, plasma composition, 68
—, resistance to dehydration, 71
—, ureogenesis in, 67
Anabolisms, cellular, starting from low molecular precursors, 39
Anaerobiosis, 34
Analogous, definition, 8
Analogy, 9
Anatomia animata, 2
Anatomic philosophy, 1
Anatomy, comparative, 2
—, general formulation, 1
Angiosperms, development, 111
Animal phylogeny, and biochemistry, 3
Apogon, 8
Apterygotes, hemolymph, inorganic cations in, 94, 104
Aragonite crystals, 141
Aragonite lamellae, in mother-of-pearl, 138, 139
Arenicola, hemoglobin in, Bohr effect of, 23
Arginase, 63, 64, 76
—, de novo arginine synthesis, 82
Arginine, biosynthesis de novo, 63, 82
—, —, from ammonia, 83
—, —, in ureogenesis, 81
—, in chlorocruorin, 13
—, in hemoglobin, 13
Arginine–vasopressin, amino acid sequence, 12
Argininosuccinate lyase, 64, 77
Argininosuccinate synthetase, 64, 77
Arthropod molting, chitinolytic system, 163
Arthropoda, chitin biosynthesis in, 54, 60

Arthropoda, *(continuation)*
—, chitinolytic systems, 60
Articulata, chitin in, 52
Artiodactyls, phylogenic relations, 26
Ascorbic acid, in insect hemolymph, 115
Asparagine, ammonia fixation, 63
Astaxanthin, carotenoid in crustaceans, 36
ATP, in biosynthesis, 34
—, isologous in cells, 7
Atrina (Pinna) nigra Durh., phase-contrast microscopy, 148
—, sheath fragments, 149
Auxotrophic mutants of microorganisms, use in study of biosynthetic pathways, 34
Avena, xanthophyll in, 37

Beauveria bassiana, chitinolytic system, 57
Bees, metamorphosis, Na/K ratio, 111
Benzoic acid, homology, 45
Bile acids, 48
—, homologous in vertebrates, 7
—, pathway, 47
Biochemical systems, dynamics of, 4
—, unity of, 3, 4
Biochemistry, and animal phylogeny, 3
—, comparative, *see* Comparative biochemistry
Bioluminescence, 163
Biosynthesis of enzymes, 5
— and phylogeny, 34–48
Biosynthetic economy, 46
Blood-ammonia level, 62
Blood clotting, 26
Blood, of insects, departure from seawater-like composition, 90
Bombyx mori, chitinolytic enzymes in, 59, 60
Bufo bufo, dehydrated, osmotic pressure and plasma composition, 69
—, in fresh water, osmotic pressure and plasma composition, 69
—, osmotic pressure and plasma composition, 68

Bufo viridis, osmotic pressure and plasma composition, 68

Calcification, of chitinous structures, 54
Calcitostracum, 139
Carbamoyl aspartate, 63, 64
Carbamoyl phosphate, 64
—, metabolic interrelationship, 63
—, synthetase, 64, 83
Carbohydrates of insect hemolymph, 115
Carbohydrate metabolism, 34
—, of insects, 115–129
Carotene, absorption in the intestine, 36
Carotenoid(s), absorption of, 36
—, of food, and retinal pigments, 160
—, structure, connection with the function of photoreception, 37
Catalase, 6, 22
—, isology, 6
Cationic patterns, in insect hemolymph, 103–111
Cells, biochemical similarity, 4
Cell theory, 5
—, Schwann, 1, 5
Cellular continuity, 5
Cephalothoracic shield, 59
Chelonians, excretion, 65
—, urea excretion, 72
Chemotaxonomy of plants, 48
Chironomus thummi, hemoglobins in blood of, 26
Chitin(s), 50–60
—, biosynthesis, 51, 53
—, in coleopterous wing-sheaths of Eocene, 133
—, crystallographic differences, 53
—, detection and determination, 51
—, in digestion, 55, 57
—, distribution, 55
—, eaters, chitinase biosynthesis, 56
—, enzymatic hydrolysis, 50
—, "free", 51, 59
—, functions in various organisms, 55

Chitin(s), *(continuation)*
—, hydrolysis products, 59
—, "masked", 51
—, mild acid hydrolysis, 50
—, in nacroin residue of mother-of-pearl, 136
—, phylogenic relations, 52
—, phylogeny, enzymapheresis, 54
— synthetase, 53
— —, change of nature of a molecular unit, 161
— —, responsible for chitin biosynthesis, 54
Chitinase(s), 50, 51, 56, 163
— biosynthesis, in the digestive tube, 57
—, —, during molting cycle, 58, 59
—, concentration, during molting cycle, 58
—, cyclic biosynthesis, by *Bombyx mori* epidermis, 60
Chitinolysis, 50–60
Chitinolytic enzymes, digestion of food, in invertebrates, 163
— —, role in parasitism, 163
— system, enzymapheresis, 56
Chitobiase, 50, 56
— biosynthesis, in the digestive tube, 57
— —, during molting cycle, 58, 59
— concentration, during molting cycle, 58
—, kinetics of, 51
Chitobiose, 50, 53
— acetamidodeoxyglycohydrolase, 50
Chitotriose, 50
Chlorhemians, 12
Chloride, in hemolymph of insects, 90
Chlorocruorin, blood pigment in sabelliforms, 12
—, homology with hemoglobin, 13
—, isoelectric point, 13
—, isology to hemoglobin, 12
—, mol. wt., 13
—, phylogeny of, 11
Chlorocruoroheme, 12

Chlorophyll, biosynthesis, 39
Cholesterol, biosynthesis, pathways of, 38
—, homology, 45
Chordata, chitin in, 52
Chromophores, 41
Chymotrypsin, homology with trypsin, 161
Citrulline, replacement of arginine, in chick, 82
Cnidaria, gastroderm, synthesis of chitinolytic enzymes, 56
Coleoptera, cationic pattern, in hemolymph, 101, 105
Colubridae proteroglyphae, digestive secretions, 163
Comparative anatomy, 2
— biochemistry, 2, 3
— —, basic concepts, 6
— physiology, 2
Conchiolin(s), lace-like reticulated sheets, 143
— membrane from *Nautilus*, 145
—, mother-of-pearl, amino acid composition, 144
—, —, from Eocene, 141
—, —, structure patterns, 136, 137
— of nacre, 136
— —, structure, 138
—, organic residue of decalcification of a molluscan shell, 135
—, in Pleistocene fossil shell of *Mercenaria mercenaria*, 133
—, prism, 141
—, —, in molluscan shells, submicroscopic structure, 145
—, prism and nacre, comparison, 146
— remnants, peptide-linkage identification, 152
Continuity, cellular, 5
Convergence, 9
—, biochemical examples of, 46
— of isologous molecules, 159
Coprostanic acids, 47
Corticotropins, amino acid sequence, 17
Cryptocephalae, 12

Cryptocephalic annelids, 12
Cypridina, 8
Cystine, in chlorocruorin, 13
—, in hemoglobin, 13
Cytochrome(s), 6, 22
—, isology, 6
— oxidase, in insects, 122
—, primary structures, comparison, 158
Cytochrome *c*, amino acid sequence, 6, 11
—, —, differences in, 31
—, animal, 26
—, convergence, 31
—, primary structure, 7
—, yeast, 26

Deaminations, 62
Dehydration, and increased uremia, 69
Dehydrogenases, change of amino acid sequence, 161
Dehydrogenation steps, in metabolic processes, 3
Deilephila euphorbiae, diapausing nymphs, amino acid concentration in hemolymph, 126
Deuterostomia coelomata, chitin in, 52
Diapause, end of, 127
Digestive secretions, of snakes, 162
DNA, bases, sequences, 6
—, base substitution in, amino acid changes in artiodactyl fibrinopeptide A, 30
—, information carried by, 15
—, hereditary information transfer, 4
—, sequence of purine and pyrimidine bases, 5

Ecdysis, 58
Ecdysone, circulation, 129
Embryonic life, water availability, ureogenesis, 65
Endoplasmic reticulum, biochemical similarity, 4
Endopterygotes—Oligoneoptera, hemolymph, inorganic cations in, 97–102

Enzymapheresis, of the chitinolytic system, 56
Enzymes, biosynthesis, 5
Episemantides, 157
—, indirect homology, 159
Evolution, diversity of organisms, 1
Evolving organisms and molecules, 157–164
Excretion synthesis, disposing of ammonia, 62
Exopterygotes—Palaeoptera, hemolymph, inorganic cations in, 94, 104
Exopterygotes—Paraneoptera, cationic pattern of hemolymph, 96, 105
Exopterygotes—Polyneoptera, hemolymph, inorganic cations in, 95–97, 105
Exuviation, 58, 59

Fibrinogen, association into fibrin, 29
Fibrinopeptides, amino acid sequences, 26
Fibrinopeptides A, amino acid sequence, 27
—, artiodactyl, amino acid changes in, 30
Fibrinopeptides B, amino acid sequence, 28
—, stepwise amino acid substitution, in artiodactyls, 29
Fireflies, 8
Flavonoid pathways, 42, 43
Fossil(s), amino acid composition of prism proteins, comparison with modern prisms, 150
—, free amino acids, 133
— remnants, insoluble, chromatography of hydrolysate, 151
— shells, isolation of individual prisms, 147
Frog skin, electrical potential difference, 67
Fructose, in hemolymph of insects, 116

Galago crassicaudatus, amino acid composition of tryptic peptides from β-chains of hemoglobin, 24

Gastric mucosa, production of chitinase, 58

Genetic unity, 1

Gluconeogenesis, biosynthetic pathway, 121

—, in insects, 117

Glucose, in hemolymph of insects, 116

Glumitocin, structure, 18

Glutamine, ammonia fixation, 63

α-Glycerophosphate accumulation, in insects, 122

— dehydrogenase, 121

Glycine, conjugation of bile acids with, in mammals, 47

—, use for detoxication, in uricotelic vertebrates, 82

Glycogen phosphorylase, activator, 120

Glycolysis, 34

—, biocatalysts in, 7

—, enzymatic step to triose phosphates, 48

— enzymes, 15

— —, during mutation, 16

—, homology, 48

Goat, fibrinopeptides, amino acid sequence, 27, 28

Gonyaulax polyhedra, 8

Graptolites, hydrolysates, amino acids in, 151–153

Graptoloidea, fossil analyses, 147

Green sulfur bacteria, metabolism, 3

Growth hormone, properties associated with specific antigenic reactions, 16

Helix pomatia, arginase in, 76

—, nitrogen metabolism, 78

Hematin catalysts, 22

Heme derivatives, isology, 6

Heme protein, reversible oxygenation, 20

Heme–heme interaction, 21

Hemoglobin, 6

Hemoglobin, *(continuation)*

—, amino acid sequence, 11

—, biosynthesis in animals, 42

—, in blood of spioniforms, 12

—, evolutionary trends, 22

— genes, evolution, 30

—, homology with chlorocruorin, 13

—, human, controlling genes of, 19

—, isoelectric point, 13

—, isology, 6

—, mammalian, 23

—, mol. wt., 13

— of monkeys, 23

— phylogeny, 21 *ff.*

—, polypeptide chains, evolution, 20

—, primary structure, 7

—, —, comparison, 158

— of primates, 23

—, primitive, 23

Hemoglobin A (adult), 20

Hemoglobin A_2, 20, 21

Hemoglobin F (fetal), 20

Hemolymph, in insects, 89

—, —, carbohydrates of, 115

—, —, fermentable sugars in, 115, 116

—, —, food and cationic composition, 108, 109

—, —, inorganic cations in, 94–102

—, —, osmotic pressure, 90

— osmolar effectors, biochemical evolution, 91–93

—, in pterygote insects, osmolar concentration, 91

Hepatopancreas, homogenates, urea production *in vitro*, 77

Hereditary information transfer, role of DNA in, 4

Heterotypic expressions, concept of, 162

Histidine, in chlorocruorin, 13

—, in hemoglobin, 13

Histolysis and the hemolymph of insects, 90

Homologous enzymes, 43, 45, 160

Homologous macromolecules, comparison along the phyletic tree, 157

Homology, 2, 9, 157
— and biosynthesis, 43
—, definition, 7
—, direct, 53
—, evolutionary aspects, 164
—, indirect, 53
— of molecules, and enzymes
    catalyzing the biosynthesis, 47
Hormonal regulation, 164
—, in insects, 121
Horse-heart cytochrome c, 23
Human β-chain, amino acid compo-
    sition, 24
Human fibrinopeptides, amino acid
    sequence, 27, 28
Hyaluronidase, in snake venoms, 163
Hydrogenation steps, in metabolic
    processes, 3
Hydrolases, of snake venoms, 163
5-Hydroquinic acid pathway, 43, 44
Hylobates lar, amino acid compo-
    sition of tryptic peptides from
    β-chains of hemoglobin, 24
Hymenoptera, cationic pattern in
    hemolymph, 106

Inoceramus sp., elongated sheath
    shred, 149
Insect(s), blood volume variability, 111
—, carbohydrate metabolism, 115–
    129
—, food and regulatory processes, 110
— hemolymph, fermentable sugars
    in, 116
—, ingestion–excretion equilibrium,
    107
— phylogeny, osmotic effectors in,
    89–112
—, with phytophagous habits, 111
—, regulation of inorganic constitu-
    ents of hemolymph, 111
Insulin, amino acid sequence, 11
—, determining sequence of bases of
    pig and sperm whale, 6
—, primary structure, 7
—, —, comparison, 158
—, structure changes, during infor-
    mation transfer, 15

Irones, 36
Isologues, definition, 6
Isology, 9
— and biosynthesis, 43
—, degree of, in various macro-
    molecules, 7
Isoprenoids, 36
Isotocin, amino acid sequence, 18

α-Keratin, in paleoproteins, 134
α-Ketonic acids, in insect hemo-
    lymph, 115
Krebs cycle, 34

Lampetra fluviatilis, hemoglobin of,
    26
Lemur catta, amino acid composition
    of tryptic peptides from β-chains
    of hemoglobin, 24
Lemur fulvus, amino acid compo-
    sition of tryptic peptides from
    β-chains of hemoglobin, 24
Lemur variegatus, amino acid compo-
    sition of tryptic peptides from
    β-chains of hemoglobin, 24
Lepidoptera, amino acid concen-
    tration in hemolymph, 125
—, cationic pattern in hemolymph,
    105
—, evolution, 110
Luciferins, analogy, 8
Luminous bacteria, 8
— organisms, selected examples, 8
Lynen's cycle, 34
Lysine, in chlorocruorin, 13
—, in hemoglobin, 13
—, homology, 45
Lysine–vasopressin, amino acid
    sequence, 18
Lysozyme, primary structure, 13, 15

Macromolecules, phylogeny of, 157
Melanocyte-stimulating hormone,
    see MSH
Melanotropic hormone, amino acid
    sequence, 11
—, primary structure, 7
Mesobilierythrin, 41

Mesobiliviolin, 41

Mesotocin, amino acid sequence, 18

Messenger RNA, in information transfer, 15

Metabolic pathways, survey of, 42

Metabolism, units of, 5

Metamorphosis, bees, Na/K ratio, 111

—, change of cationic composition of hemolymph during, 104

—, wasps, Na/K ratio, 111

Mevalonic acid, in biosynthesis of terpenes, 36

Mevalonic pathway, metabolic sequences, 46

Mitochondria, biochemical similarity, 4

Molecular biochemical units, 4

Molecular evolution, and protein primary structure, 164

Mollusks, shells, chitin in, 55

Molting hormones, 58

Mother-of-pearl, structure, 138

MSH, amino acid sequence, 17

—, structural relation with ACTH, 16

Mucopolysaccharide fibrils, in conchiolin of nacre, 138

Mutation, spectrum of proteins during, 16

Nacre(s), amino acid composition, 147

— and prism conchiolins, comparison, 146

— residue, chromatography, 142

—, structure patterns, 137, 138

Nacreous conchiolin, amino acid pattern, 141

— layers of shells, electronmicroscopy, 145

— organic remnants in fossils, electron micrographs, 143

Nacrine, soluble and insoluble, 136

Nacroin fibers, of *Pinctada margaritifera*, 135

—, presence of chitin in, 136

Nephridium excreta, *Helix pomatia*, nitrogen compounds in, 80

Nephridium, *Helix pomatia*, nitrogen compounds in, 80

Neurohypophyseal peptides, structure, 18

—, in vertebrates, hypothetical phylogeny, 19

Neurohypophysis hormones, amino acid sequence, 18

Neuromuscular transmission, in insects, 110

Neurospora crassa, chitin biosynthesis, 53

Nitrogen metabolism, terminal products of, 62–84

Nucleic acids, homologous macromolecules, comparison, 158

—, isologous base sequences, 7

—, primary sequences, 159

—, primary structures, 160

—, —, degree of isology, 157

Nucleosides, breakdown to uric acid, 75

Odontosyllis, 8

Oligoneoptera, cationic pattern in hemolymph, 105

Omphali flavida, 8

Ontogenic development, 65

Ornithine carbamoyltransferase, 64

—, conjugation in birds, 83

— cycle, 83

—, replacement of arginine in diet of insects, 82

Ornithuric acid synthesis, 83

Osmoregulation, and ureogenesis, 83

Osmotic deficit, 68

Osmotic effects of blood components in insects, 92

Osmotic effectors in insect phylogeny, 89–112

Osmotic pressure regulation, in insects, 89

—, in insect hemolymph, 112

Oviparous reproduction, in elasmobranchii, 66

Ox, fibrinopeptides, amino acid sequence, 27, 28

Oxidative phosphorylation, biocatalysts in, 7
Oxygen carriers, analogy, 8
Oxytocin, amino acid sequence, 18
—, in vertebrates, activity in reproduction, 19

Paleobiochemistry, 133
Paleoproteins, 133–154
—, primary structure, 154
Panorpoid complex, divergence of,
103
Papio doguera, amino acid compo-
sition of tryptic peptides from
β-chains of hemoglobin, 24
Parallelism, 46
Porphyrins from coprolites of
Eocene crocodiles, 133
Pentose cycle, 163
Peptides, phylogeny, 11–31
Peridermic structures, use of chitin
for, 53
Perodicticus potto, amino acid com-
position of tryptic peptides from
β-chain of hemoglobin, 24
Peroxidase, 6, 22
—, isology, 6
Phenomenological aspects, 2
Pholas dactylus, 8
Phosphagen synthesis, 64
Photoreception, connection of caro-
tenoid structure with, 37
— system in vertebrates, 37, 39
Photoreceptors, 37
Photosynthesis, 3
— pigments, 41
Phycomyces sporangiophores,
β-carotene in, 37
Phylogenic tree, molecular changes
along, 4
Phylogeny and systematics, 3
Physiological radiation, 162
—, definition, 54
Physiology, comparative, 2
—, general, definition, 1
Phytophagous insects, 107–109
Pig, fibrinopeptides, amino acid
sequence, 27, 28

Pinna affinis, prism sheath facets, 149
—, tubular prism sheaths, 148
Pitocin, action on mammalian uterus,
164
Poly-β-1,4-(2-acetamido-2-deoxy)-D-
glucoside glycanohydrolases, 50
Polypeptide chains in hemoglobins,
evolution, 20
Polypeptide fibrils in conchiolin of
nacre, 138
Porphyrins, biosynthesis, 39, 41
—, fossil, from sedimentary rocks,
141
—, homology, 45
— pathway, 40
Porphyropsin system, 39
Prisms, amino acid composition, 147
—, modern, amino acid composition,
comparison with fossils, 150
—, and nacre conchiolins, comparison, 142, 146
Prodigiosin, biosynthesis, 45, 46
Prolactin, 164
Propithecus verreauxi, amino acid
composition of tryptic peptides
from β-chains of hemoglobin, 24
Protein(s), amino acid sequences, and
homology, 157
—, analogy, 8
—, catalysts, and homology, 157
—, degradation, nitrogen end-
product, 62
—, homologous macromolecules,
comparison, 158
—, isologous primary structures, 7
— macromolecules, primary structure, 6
—, new, appearance during evolution,
16
—, phylogeny, 11–31
—, primary structures, 159, 160
—, —, and molecular evolution, 164
— remains, in fossil shells, 133
—, specificity of species, 16
—, structure(s), changes, indirect
result from change in DNA
structure, 19
—, —, hierarchy of, 16

Protein(s), structure(s), *(continuation)*
—, —, primary, concepts of isology, homology, analogy and convergence, 9
— synthesis, 64
Protein–chitin complex in paleoproteins, 134
Proteolytic enzymes, 58
Protostomia, chitin in, 52
—, chitin biosynthesis in, 53
Pseudo-nacre, 139
Pterygote insects, osmolar concentration of hemolymph, 91
Purine(s), bases, sequence, 5
—, biosynthesis, 72, 73
—, catabolism, end-point, 76
—, metabolism, 75
—, ring synthesis, Buchanan's scheme, 74
—, synthesis, and uric acid excretion, 83
Purine-containing nucleotides, biosynthesis, 72, 73
Purinolysis, 75, 76, 84
Pyrimidine bases, sequence, 5
Pyrimidine synthesis, metabolic interrelationships, 63, 64

Rabbit, fibrinopeptides, amino acid sequence, 27, 28
Radioactive isotopes, use in study of biosynthetic pathways, 34
Rana cancrivora, ecology, 67
Rana temporaria, chitin digestion, 58
—, dehydrated, osmotic pressure and plasma composition, 69
—, in fresh water, osmotic pressure and plasma composition, 69
—, osmotic pressure and plasma composition, 68
Reindeer, fibrinopeptides, amino acid sequence, 27, 28
Renilla reniformis, 8
Reptiles, adaptation to terrestrial life, 71
—, permeability of skin, 72
Respiratory pigment turnover, rise of, at end of diapause, 127

Retinal pigments, 159, 160
Retinene I, 37
Retinene II, 39
Retinenes, zoological distribution, 159, 160
Rhodopsin, 37
— system, 39
Ribonuclease, amino acid sequence, 11
—, pancreatic, amino acid sequence, 14
—, —, primary structure, 13, 14
—, primary structure, 7
—, push-pull effect, 15
—, and tertiary structure of protein, 161
—, topography of active site, 15
RNA, phosphate ester bond, cleavage, 15

Sabellariides, 12
Sabellidae, 12
Sabelliforms, 12
Saturnia pavonia, diapausing nymphs, amino acid concentration in hemolymph, 127
Scymnol, 47
Sea turtle, adaptation to fresh water, 72
Semantides, 157
Serpulidae, 12
Serum albumin, during mutation, 16
—, human beings deprived of, 15
Sheep, fibrinopeptides, amino acid sequence, 27, 28
Sibling species, variation of genotype, 158
Silicification, use of chitin in, 54
Silk fibroin, 118
Silk glands, 119
Silk worm, ontogenesis of, 119
Similarity, from cell to cell, 5
Somatotropins, structure changes, during information transfer, 15
Sphinx ligustri, free amino acids, 123, 124
—, nymphs, amino acid concentration, in hemolymph, 128

Spongillidae, use of chitin by, 53
Sporangiophores of *Phycomyces*, β-carotene in, 37
Steroid hormones, biosynthesis, 48
— pathway, 35, 36, 46
— —, metabolic sequences, 46
Streptomyces antibioticus, chitinase of, 51
Structure, units of, 5
Sugars, fermentable, in insect hemolymph, 115, 116
Systematics, and phylogeny, 3

Terpenoid pathway, 35, 36, 46
Testudo hermanni, plasma composition, monthly variations, 70
Tetrahymena, ureogenesis in, 83
Thrombin, 26
Tortoise, plasma composition, monthly variations, 70
Trehalase activity in hemolymph of *Bombyx mori*, 118
Trehalase, in insect hemolymph, 117
Trehalose, biosynthesis from glucose, 119
—, biosynthetic pathway, 120
—, concentration, in hemolymph of *Bombyx mori*, 118
— in hemolymph of insects, 115, 116
— phosphatase, 120
—, synthesis by insects, 119
Trehalose-phosphate–UDP glucosyltransferase, 120
Trehalosemia, 120
—, in insects, 118
Trypsin, homology with chymotrypsin, 161
Tryptic peptides, primate, amino acid composition, 24, 25
Tupaia glis, hemoglobin of, 23
Tyrosine, in insect hemolymph, 115
—, in prism conchiolin, 145

Units, molecular biochemical, 4
Unity of biochemical plan, 5
Unity, systematic search for, 1
Urea, biosynthesis, disposal of ammonia, 63

Urea, *(continuation)*
—, concentration in blood, increase of, 69
— cycle enzymes, 81
— — —, in liver of amphibians, 65
— — —, vertebrates, phylogeny, 65
— —, metabolic interrelationships, 63
—, excretion increase, 81
— nitrogen, in urine, amount of, 77
— retention, 66
—, synthesis *de novo*, 82
— —, and osmotic problems, 66
Urease, 74, 75, 83
Urechis hemoglobin, 22
Uremia, physiological, 69
Ureogenesis, biochemical system used for, 63
—, in birds, 71
—, connected with osmoregulation, 83
—, *de novo* arginine synthesis, 81, 82
—, enzymes catalyzing, 64
—, evolution of, 64
—, in Protozoa, 83
—, in reptiles, 71
—, steps of, 64
Ureotelic nitrogen metabolism, examples among invertebrates, 81
— vertebrates, 63
— —, liver of, 64
— —, ureogenesis, 82
Uric acid, biosynthesis in birds, 73, 74
—, — in hepatopancreas, 72
—, excretion, and purine synthesis, 83
—, formation in animals, by degradation of purine-containing nucleotides, 72
—, in insect hemolymph, 115
Uricase, 75
Uricolysis, 75
Uricolytic enzyme system, 76
Uricotelic nitrogen metabolism, 81
— vertebrates, ecological adaptation, 76

Uridine-diphosphate–acetylglucosamine, acetylglucosamine donor
in chitin biosynthesis, 53

Vasopressin, in vertebrates, hydromineral regulation, 19
Vasotocin, amino acid sequence, 18
Vertebrates, cytochrome *c*, amino
acid sequence, 6
—, evolutionary tree, 66
—, nitrogen metabolism, terminal
products, 81
Violability, of protein structures, 16

Viperidae, digestive secretions, 163
Vitamin A, 37

Wasps, metamorphosis, Na/K ratio,
111
Wiener's scheme, uric acid synthesis
in birds, 73, 74

Xanthine oxidase, 75
Xanthophylls, absorption of, 37

Yeast cytochrome *c*, amino acid
sequence, 6